ACOUSTICS RESEARCH AND TECHNOLOGY

NOISE REDUCTION

METHODS, APPLICATIONS AND TECHNOLOGY

Acoustics Research and Technology

Additional books in this series can be found on Nova's website under the Series tab.

Additional e-books in this series can be found on Nova's website under the e-book tab.

ACOUSTICS RESEARCH AND TECHNOLOGY

NOISE REDUCTION

METHODS, APPLICATIONS AND TECHNOLOGY

MARC FARROW
EDITOR

nova
science publishers
New York

Library of Congress Cataloging-in-Publication Data

ISBN: 978-1-53613-541-1

Published by Nova Science Publishers, Inc. † New York

CONTENTS

PREFACE

In *Noise Reduction: Methods, Applications and Technology*, the authors open with an overview of image denoising approaches. The review covers additive and multiplicative noise reduction methods in the spatial, frequency and wavelet domains, as well as their applications. Finally, experimental qualitative and quantitative evaluations on synthetic and real images corrupted by noise of different distributions are presented and discussed. Then next chapter begins with introduction to multiple camera systems: their usage, structure, hardware and applications. Some examples of existing systems are be presented along with quantitative comparison of parameters like: sensor resolution, technology, frame rate, shutter-related features, need and means for temporal synchronization. Next, optical phase induced interference noise (PIIN) in a sensor network is reduced by using optical phase modulation in both a SCOS and a fiber Bragg grating (FBG), and strain induced noise on the optical fiber sensor is reduced using a differential sensor setup called a push-pull SCOS. Phase modulation and differentiation are then combined to allow for simultaneous noise reduction from different noise sources and allow for optical sensing of electric fields in harsh environments. Additionally, the signal processing approaches that exploit the spatial and temporal-spectral characteristics of the flaw echo and noise for enhanced defect detection and noise reduction in ultrasonic NDE are presented. Results from simulation studies and experiments on the real industrial samples are presented to evaluate the techniques, and demonstrate their robustness and advantages over the classical methods. Lastly, the authors show how Riga City Municipality has managed to fulfil the main goal of the 'Directive 2002/49/EC of the European Parliament and of the Council of 25 June 2002 relating to the assessment and management of environmental noise' – to reduce the number of environmental noise affected people in Riga agglomeration.

Chapter 1 - Noise is one of the most widespread problems in imaging applications such as computer vision, computer graphics, pattern recognition, image restoration, editing and completion, image retrieval and image understanding. In the past decades, several algorithms have been proposed for image denoising. In spite of the sophistication of the recently proposed algorithms, denoising methods share the same principle; reduce noise effect while preserving the informative structures of the underlying image. Most of the existing methods have shown their efficiency in different situations. However, failure cases still exist and some of the proposed approaches have not yet attained a desirable level of applicability. The purpose of this chapter is to give an overview of image denoising approaches. The review covers additive and multiplicative noise reduction methods in the spatial, frequency and wavelet domains, as well as their applications. Finally, experimental qualitative and quantitative evaluations on synthetic and real images corrupted by noise of different distributions are presented and discussed.

Chapter 2 - The chapter begins with an introduction to multiple camera systems: their usage, structure, hardware and applications. Some examples of existing systems are being presented along with quantitative comparisons of parameters like: sensor resolution, technology, frame rate, shutter-related features, need and means for temporal synchronization. Next, means for noise estimation in multiview systems are presented. This part covers references to the estimation of noise parameters (statistical distributions, noise independence etc.) and also examples of measurements with a discussion of the presented data and caveats. The third part of the chapter presents considerations about how the knowledge about noise existing in multiview camera systems can augment video processing used in applications in the field. Among other mentioned applications, including light-field video acquisition and compression, multiview video plus depth (MVD) acquisition and compression and also stereoscopic depth estimation. The chapter focuses on two subjects: stereoscopic depth estimation and compression. On these examples, some ideas about improving existing algorithms are shown along with some practical results that can be attained. In the case of stereoscopic depth estimation, the relation between noise in the camera system and the matching technique is shown on the example of depth estimation technique based on Maximum A posteriori Probabiity (MAP) rule. In the case of compression, it is shown how noise reduction and extraction as a preprocessing step can be used in order to improve compression efficiency and increase the subjective quality of multiview video.

Chapter 3 - Test and evaluation needs are becoming more and more demanding on the sensors being used. Fiber optic sensors are known for their ability to survive in harsh environments such as high pressure and temperature like those within oil/gas wells, or high vibration like those with electromagnetic railguns. As these sensors become commercially available, one challenge is overcoming multiple noise sources simultaneously in the sensor and its network imposed from the harsh environment. Optical amplifiers are typically used in fiber optic sensors to improve the signal-to-noise ratio (SNR). This method is suitable for boosting the signal above the floor caused by electrical noise. However, certain noise sources in fiber optic sensors directly modulate the optical signal. This can be due to other measures and being induced on the system. For example, while a slab coupled optical sensor (SCOS) is sensitive to electric field, it is also inherently sensitive to strain; therefore, any strain noise imposed on the sensor would overwhelm the electric field measurement and be considered noise. In this chapter, optical phase induced interference noise (PIIN) in a sensor network is reduced by using optical phase modulation in both a SCOS and a fiber Bragg grating (FBG). Strain induced noise on the optical fiber sensor is reduced using a differential sensor setup called a push-pull SCOS. Phase modulation and differentiation are then combined to allow for simultaneous noise reduction from different noise sources and allow for optical sensing of electric fields in harsh environments. These principles are applicable to other types of fiber optic sensors and sensors in general.

Chapter 4 - Reliable and robust defect detection is a challenging yet essential problem in ultrasonic non-destructive evaluation (NDE). The flaw echoes are usually contaminated by high-level time-invariant grain noise originating from the material microstructures. This phenomenon becomes even worse when inspecting coarse-grained materials like austenitic stainless steel, alloy, carbon-reinforced composite and concrete, which however, form some of the most important and commonly used industrial materials. This problem has attracted great attentions from the NDE community and a wide variety of techniques have been investigated to enhance the signal-to-noise ratio (SNR) and flaw detection. In this chapter, the signal processing approaches that exploit the spatial and temporal-spectral characteristics of the flaw echo and noise for enhanced defect detection and noise reduction in ultrasonic NDE are presented. The first category of techniques exploits the spatial diversity created by the transducer array, and utilises spatial filtering to discriminate the flaw echo from the clutter noise. Particularly, minimum variance beamforming seeks to minimize the power of the array output while maintaining a unit gain on the focal point, and weighs the observations with data-dependent apodization. The second category of approaches investigates the time-frequency analysis and statistical properties of flaw echoes and grain noise. In particular, the advanced matched filter is designed to match unknown flaw echoes rather than the transmitted signal, and utilise an optimization paradigm to tweak and search the best filter parameters. Results from simulation studies and experiments on the real industrial samples are presented to evaluate the techniques, and demonstrate their robustness and advantages over the classical methods.

Chapter 5 - The purpose of this chapter is to show how Riga City Municipality has managed to fulfil the main goal of the 'Directive 2002/49/EC of the European Parliament and of the Council of 25 June 2002 relating to the assessment and management of environmental noise' – to reduce the number of environmental noise affected people in the Riga agglomeration. According to Directive 2002/49/EC, for this purpose the

Riga City Municipality drew up the 'Action Plan for Reduction of Environmental Noise in Riga Agglomeration' in 2014, which has been approved by the Riga City Council on July 8, 2014. This Action Plan scheduled the measures to reduce or to prevent the harmful effects of environmental noise emitted by road traffic, especially on major roads, by rail traffic and airplanes, as well as by industrial sites and open air industrial equipment. As the Directive 2002/49/EC obligates the review and to revise Action Plans at intervals of at least every five years, Riga City Municipality timely carried out the efforts to assess the data about noise affected people as well as gains and possible losses after the 'Action Plan for Reduction of Environmental Noise in Riga Agglomeration' (2014) was implemented. The Action Plan contains the medium-term and long-term activities devoted to the reduction of noise pollution in the dwelling areas, in public parks and other quiet areas in Riga City. All planned measures belong to noise reduction technical solutions – sound barriers (hedgerows and tree rows included), city transport infrastructure pavement and tram railway quality improvement, cargo train speed limitation on inner city railways and building façade sound insulation ability improvement by glass panels as well. It was estimated to spend some 90,000 EUR for medium-term measures but for long-term measures - more than 10 million EUR.

In: Noise Reduction
Editor: Marc Farrow
ISBN: 978-1-53613-541-1

Chapter 1

DENOISING OF DIGITAL IMAGES: METHODS AND APPLICATIONS

Adib Akl*[*] *and Charles Yaacoub
Faculty of Engineering,
Holy Spirit University of Kaslik (USEK),
Jounieh, Lebanon

ABSTRACT

Noise is one of the most widespread problems in imaging applications such as computer vision, computer graphics, pattern recognition, image restoration, editing and completion, image retrieval and image understanding. In the past decades, several algorithms have been proposed for image denoising. In spite of the sophistication of the recently proposed algorithms, denoising methods share the same principle; reduce noise effect while preserving the informative structures of the underlying image. Most of the existing methods have shown their efficiency in different situations. However, failure cases still exist and some of the proposed approaches have not yet attained a desirable level of applicability.

The purpose of this chapter is to give an overview of image denoising approaches. The review covers additive and multiplicative noise reduction

[*] Corresponding Author Email: adibakl@usek.edu.lb.

methods in the spatial, frequency and wavelet domains, as well as their applications. Finally, experimental qualitative and quantitative evaluations on synthetic and real images corrupted by noise of different distributions are presented and discussed.

Keywords: denoising, filtering, noise, restoration, speckle, wavelets

1. Introduction

Noise is a common problem that has always existed in the field of digital image processing. It is found in different imaging applications such as medical images, remote sensing, sonar and radar imaging, computer graphics, computer vision, pattern recognition, image restoration, image retrieval, image understanding, image editing and completion [1], etc.

Generally, noise originates from the superposition of echoes of random values (amplitudes and phases) during acquisition and/or transmission. It is considered as a process with random variables characterized by a probability density function (PDF), such as Gaussian, Rayleigh, Erlang (Gamma), exponential, uniform, etc.

Noise can be additive or multiplicative. Depending on its nature and distribution, it gives the image a mottled, grainy, textured, or snowy appearance. Its effect is naturally more significant when the objects being imaged are relatively small and have low contrast [2].

In the past decades, numerous image denoising techniques were proposed. In spite of the satisfactory results obtained with these methods in different cases and scenarios, most of them show an outstanding performance when the image model corresponds to the algorithm assumptions, but fail in difficult situations and create artifacts or remove image fine structures. Consequently, a universal "best" approach has yet to be found and the search for an efficient image denoising method is still a valid challenge.

This chapter provides a review of image denoising approaches with focus on additive and multiplicative noise reduction methods in the spatial, frequency and wavelet domains, as well as their applications. A comparison of various methods is carried out and evaluated qualitatively and quantitatively on synthetic as well as real images corrupted by noise of different distributions.

The remainder of this chapter is organized as follows. The noise formation and distributions are presented in section 2. Image denoising methods are reviewed in section 3. Experimental evaluations are discussed and analyzed in section 4. Finally, conclusions are drawn in section 5.

2. NOISE SOURCES AND DISTRIBUTIONS

2.1. Noise Formation

The principle sources of noise in digital images arise during image acquisition and/or transmission. Image noise can be produced by the sensor and circuitry of an image scanner or digital camera. It can also originate in film grain and in the unavoidable shot noise of an ideal photon detector. The performance of imaging sensors is affected by a variety of factors, such as environmental conditions during image acquisition, and by the quality of the sensing elements themselves. For instance, in acquiring images with a CCD (Charge-Coupled Device) camera, light levels and sensor temperature are major factors affecting the amount of noise in the resulting image. Images are corrupted during transmission principally due to interference in the channel used for transmission. For example, an image transmitted using a wireless network might be corrupted as a result of lightning or other atmospheric disturbance [3]. In fact, no imaging method is free of noise, but noise is much more prevalent in certain types of imaging procedures than in others.

2.2. Common Noise Probability Density Functions

As mentioned earlier, image noise is considered as a random variable characterized by its Probability Density Function (PDF). The following are among the most common PDFs found in image processing applications.

2.2.1. Gaussian Noise

Gaussian noise, also called normal noise, is used frequently in practice because of its mathematical tractability in both the spatial and frequency domains. In fact, this tractability is so convenient that it often results in Gaussian models being in situations in which they are marginally applicable at best. The PDF of a Gaussian random variable z is given by:

$$p(z) = \frac{1}{\sigma\sqrt{2\pi}} \exp\left(-\frac{(z-\mu)^2}{2\sigma^2} \right), \tag{1}$$

where μ is the mean value of z, and σ its standard deviation. Approximately 70% of z values are in the range $[\mu\text{-}\sigma, \mu+\sigma]$ and about 95% in the range $[\mu\text{-}2\sigma, \mu+2\sigma]$ [3].

2.2.2. Rayleigh Noise

The PDF of Rayleigh noise is given by:

$$p(z) = \begin{cases} \frac{2}{b}(z-a)e^{-\frac{(z-a)^2}{b}} & , \ z \geq a \\ 0 & , \ z < a \end{cases}, \tag{2}$$

where a and b are the model parameters. The mean and variance of a Rayleigh PDF are respectively given by:

$$\mu = a + \sqrt{\frac{\pi b}{4}}, \ \sigma^2 = \frac{b(4-\pi)}{4}. \tag{3}$$

2.2.3. Erlang (Gamma) Noise

The PDF of Erlang noise is given by:

$$p(z) = \begin{cases} \frac{a^b z^{b-1}}{(b-1)!} e^{-az} & , \ z \geq 0 \\ 0 & , \ z < 0 \end{cases}, \tag{4}$$

where a and b are positive integers. The mean and variance are respectively expressed as:

$$\mu = \frac{b}{a},\ \sigma^2 = \frac{b}{a^2}\ . \tag{5}$$

This noise is referred as Gamma noise when the denominator in (4) is replaced with the gamma function $\Gamma(b)$.

2.2.4. Exponential Noise

The PDF of exponential noise is expressed as:

$$p(z) = \begin{cases} ae^{-az} & ,\quad z \geq 0 \\ 0 & ,\quad z < 0 \end{cases}, \tag{6}$$

where a is a positive integer. The mean and variance are respectively given by:

$$\mu = \frac{1}{a},\ \sigma^2 = \frac{1}{a^2} \tag{7}$$

Note that the exponential distribution is a particular case of the Erlang distribution with $b = 1$.

2.2.5. Uniform Noise

The PDF of uniform noise is given by:

$$p(z) = \begin{cases} \dfrac{1}{b-a} & ,\quad a \leq z \leq b \\ 0 & ,\quad otherwise \end{cases} \tag{8}$$

The mean and variance are respectively:

$$\mu = \frac{a+b}{2},\ \sigma^2 = \frac{(b-a)^2}{12}. \tag{9}$$

2.2.6. Impulse (Salt and Pepper) Noise

The PDF of bipolar impulse noise is given by:

$$p(z) = \begin{cases} P_a & , \quad z = a \\ P_b & , \quad z = b \\ 0 & , \quad otherwise \end{cases} \tag{10}$$

If either P_a or P_b is zero, the impulse noise is called unipolar. If neither probability is zero, and especially if they are approximately equal, impulse noise will result in dark and bright dots that resemble salt and pepper granules randomly distributed over the image. For this reason, bipolar impulse noise is also known as salt-and-pepper noise.

3. IMAGE DENOISING METHODS: STATE OF THE ART

A detailed review of image denoising methods is presented in this section, covering spatial linear and nonlinear filters, adaptive and Wiener filtering, speckle noise reduction and wavelet-based filters.

3.1. Spatial Filters

Spatial filtering is generally used in the case of additive noise:

$$g(x,y) = f(x,y) + n(x,y), \tag{11}$$

where $n(x, y)$ is the noise at coordinates (x, y), $f(x, y)$ the original image and $g(x, y)$ the noisy image. The spatial restoration filter then leads to a filtered output image $z(x, y)$ as illustrated in Figure 1.

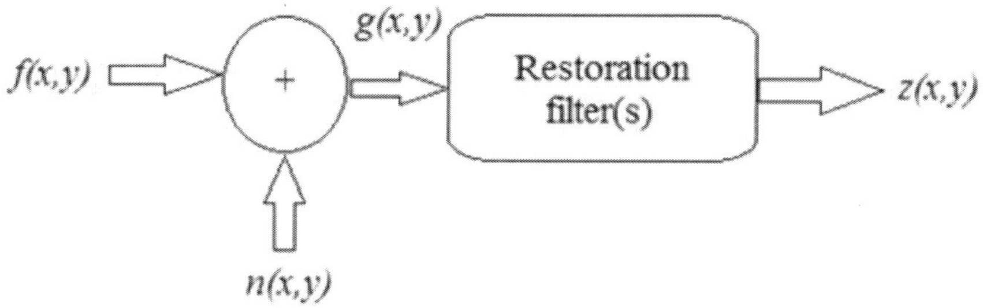

Figure 1. Model of additive noise formation and restoration process.

3.1.1. Spatial Mean Filters

Spatial mean filters smooth the local variations of the corrupted image. In other words, noise is reduced by blurring [3].

3.1.1.1. Arithmetic Mean Filter

The arithmetic mean filter computes the average of pixel values as follows:

$$z(x, y) = \frac{1}{mn} \sum_{(s,t) \in S_{xy}} g(s,t) \; , \tag{12}$$

where S_{xy} represents the set of coordinates in a rectangular window (neighborhood) of size $m \times n$, centered at pixel position (x, y).

This filter computes the average value of the corrupted image $g(x,y)$ in the area defined by S_{xy}. The value of the restored image at point (x,y) is simply the arithmetic mean computed using the pixels in the region defined by S_{xy}. This operation can be implemented by performing image convolution with a spatial filter of size $m \times n$ in which all coefficients have a value of $1/mn$.

3.1.1.2. Geometric Mean Filter

In geometric mean filtering, each restored pixel is calculated as the product of the pixels in the sub-image window, raised to the power of $1/mn$:

$$z(x,y)=\left[\prod_{(s,t)\in S_{xy}} g(s,t)\right]^{\frac{1}{mn}} \tag{13}$$

Note that the geometric mean filter achieves more smoothing than the arithmetic mean filter. However, it tends to better preserve image details.

3.1.1.3. Harmonic Mean Filter

The output (or restored) image obtained by a harmonic mean filter is given by:

$$z(x,y)=\left[\prod_{(s,t)\in S_{xy}} g(s,t)\right]^{\frac{1}{mn}} . \tag{14}$$

Note that the harmonic mean filter works well for salt noise, but fails for pepper noise. It also suits other types of noise like Exponential and Erlang noise.

3.1.1.4. Contraharmonic Mean Filter

The contraharmonic mean filter yields a restored image using the following operation:

$$z(x,y)=\frac{\sum_{(s,t)\in S_{xy}} g(s,t)^{Q+1}}{\sum_{(s,t)\in S_{xy}} g(s,t)^{Q}} , \tag{15}$$

where Q is the order of the filter.

This filter is well suited for reducing the effects of salt-and-pepper noise; for positive values of Q, the filter eliminates pepper noise and for negative values of Q, it eliminates salt noise. If $Q = 0$, the contraharmonic mean filter reduces to the arithmetic mean filter, and it reduces to the harmonic mean filter if $Q = -1$.

3.1.2. Spatial Order-Statistic Filters

Order-statistic filters are spatial filters whose response is based on ordering (ranking) the values of the pixels contained in the image area encompassed by the filter. The response of the filter is determined by the ranking result [3].

3.1.2.1. Median Filter

The median filter is the most useful order-statistic filter. It replaces the value of a pixel by the median value in the neighborhood of that pixel. For example, in a 3x3 neighborhood, the median is the 5th larger value. This operation is given by the expression:

$$z(x,y) = \underset{(s,t)\in S_{xy}}{median} \{g(s,t)\}. \tag{16}$$

For certain types of random noise, median filters provide excellent noise reduction capabilities, with considerably less blurring than linear smoothing filters of similar size. These filters are particularly effective in the presence of both bipolar and unipolar impulse noise, and they yield excellent results for images corrupted by salt-and-pepper noise because of its appearance as white and black dots superimposed on the image.

3.1.2.2. MAX and MIN Filters

The MAX filtering of an image can be expressed as:

$$z(x,y) = \max_{(s,t)\in S_{xy}} \{g(s,t)\}. \tag{17}$$

This filter is useful for finding the brightest points in an image. As a result of the max selection process in the sub-image area S_{xy}, pepper noise is reduced since it has very low values.

Similarly, the MIN filter yields a restored image based on:

$$z(x,y) = \min_{(s,t)\in S_{xy}} \{g(s,t)\}. \tag{18}$$

This filter is useful for finding the darkest points in an image, and it also reduces salt noise as a result of the min operation.

3.1.2.3. Midpoint Filter

The midpoint filter computes the midpoint between the maximum and the minimum values in the area encompassed by the filter window:

$$z(x,y) = \frac{1}{2}\left[\max_{(s,t)\in S_{xy}} \{g(s,t)\} + \min_{(s,t)\in S_{xy}} \{g(s,t)\}\right]. \tag{19}$$

This filter combines order statistics and averaging. It works best for Gaussian and Uniform noise types.

3.1.2.4. Alpha-Trimmed Mean Filter

This filter has a parameter d. It removes the $d/2$ lowest and the $d/2$ highest intensity values of $g(s, t)$ in the neighborhood S_{xy}. $g_r(s, t)$ represents the remaining mn - d pixels. The alpha-trimmed mean filter consists of averaging these remaining pixels:

$$z(x,y) = \frac{1}{mn-d} \sum_{(s,t)\in S_{xy}} g_r(s,t), \tag{20}$$

where d can range from 0 to $mn - 1$.

The alpha-trimmed filter reduces to the arithmetic mean filter when $d = 0$ and becomes a median filter when $d = mn - 1$. For other values of d, this filter is useful in situations involving multiple types of noise, such as a combination of salt-and-pepper and Gaussian noise.

3.1.3. Spatial Adaptive Filters

The filters discussed thus far are applied to an image regardless of how image characteristics vary from one pixel to another. The adaptive filters behavior changes based on statistical characteristics of the image inside the filter region defined by the *m* x *n* rectangular window S_{xy}. Adaptive filters are capable of a superior performance compared to that of the filters discussed thus far, but the price paid for improving filtering efficiency is an increase in filter complexity.

3.1.3.1. Adaptive, Local Noise Reduction Filter

This filter is based on the means and variances since they are quantities closely related to the appearance of an image; the mean gives a measure of average intensity in the region over which the mean is computed, and the variance gives a measure of contrast in that region.

The filter operates on a local region S_{xy}. The response of the filter at any point (x,y) on which the region is centered is based on four quantities:

- $g(x, y)$, the value of the noisy image at pixel (x, y)
- σ_n^2, the variance of the noise corrupting $f(x, y)$ to form $g(x, y)$
- m_L, the local mean of the pixels in S_{xy}
- σ_L^2, the local variance of the pixels in S_{xy}

The behavior of the filter is as follows:

- If $\sigma_n^2 = 0$, the filter should return $g(x, y)$; this is the no-noise case in which $g(x, y) = f(x, y)$.
- If the local variance is high relative to σ_n^2, the filter should return a value close to $g(x, y)$ because a high local variance typically is associated with edges, and these should be preserved.

- If the two variances are almost equal, the filter should return the arithmetic mean value of the pixels in S_{xy}. This condition occurs when the local area has the same properties as the overall image, and local noise is to be reduced simply by averaging.

The adaptive expression allowing to obtain these three conditions can be written as:

$$z(x,y) = g(x,y) - \frac{\sigma_n^2}{\sigma_L^2}\left[g(x,y) - m_L\right] \tag{21}$$

Based on this equation, we need to know or estimate the variance of the overall noise and the other parameters should be computed from the pixels in S_{xy} at each location *(x, y)* on which the filter window is centered.

Note that, if $\sigma_n^2 > \sigma_L^2$, we may have negative values. One approach is to allow these values to occur, and then rescale the intensity values at the end, and the disadvantage would be a loss of dynamic range in the image.

3.1.3.2. Adaptive Median Filter

The static median filter that we discussed earlier performs well if the spatial density of the impulse noise is small ($P_a < 0.2$ and $P_b < 0.2$). The adaptive median filter is better when P_a and P_b are larger than 0.2. It also works in a rectangular window area S_{xy} whose size increases adaptively with respect to the local area statistics. We will not go into details by explaining the adaptive median algorithm and how it works because it fits best for salt-and-pepper noise, which is not our main concern in this chapter. The main benefits of the adaptive median filter over the static median filter is that it seeks to preserve details while smoothing non-impulse noise and reduces distortion, such as excessive thinning or thickening of object boundaries.

3.2. Wiener Filtering

3.2.1. Combination of Additive Noise and Degradation Function

An acquired image can be degraded by a distortion process that is not additive. For example, a degradation of an ultrasound image can be caused by the ultrasonic probes resulting from their disinfection with alcohol or by linear motion or unfocused optics.

The degradation process is modeled as a degradation function *H*, that together with an additive noise term, operates on an input image $f(x, y)$ to produce a degraded image $g(x, y)$ (as shown in the degradation and restoration model of Figure 2). If *H* is linear and position invariant, $g(x, y)$ is given in the spatial domain by:

$$g(x,y) = h(x,y) * f(x,y) + n(x,y), \tag{22}$$

where $h(x,y)$ is the spatial representation impulse response of the degradation function and $*$ is the convolution operator.

We may write the model in an equivalent frequency domain representation:

$$G(u,v) = H(u,v)F(u,v) + N(u,v), \tag{23}$$

where capital letters indicate representation in the frequency domain, and *H* is the transfer function of the degradation process.

The objective of restoration is to obtain an estimate $z(x,y)$ of the original image $f(x,y)$, given $g(x,y)$ and some information about the additive noise $n(x,y)$ and the degradation function *H* [3].

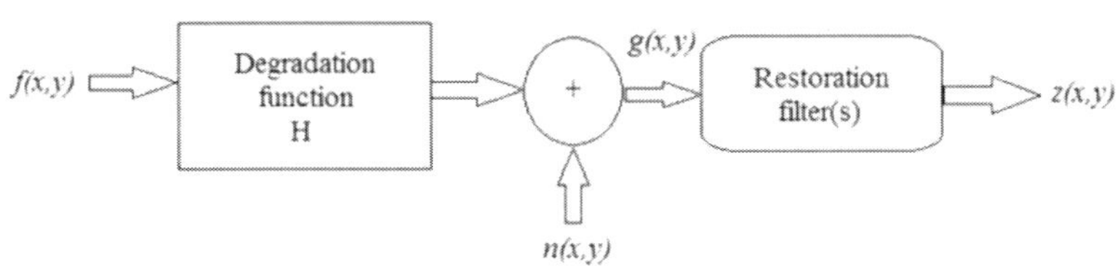

Figure 2. Model of image restoration process, after noise addition and degradation.

3.2.2. The Wiener Filter

The Wiener filter process considers images and noise as random variables. The method consists in finding an estimate z of the uncorrupted image f such that the mean square error (the mean of the squared differences between pixels) between them is minimized. For this reason, the Wiener filter is commonly referred to as the *minimum mean square error filter* or *the least square error filter*. A statistical form of the Mean Square Error (MSE) is given by:

$$e^2 = E\left\{(f-z)^2\right\} \tag{24}$$

when the restored image is very close to the uncorrupted one, the MSE is minimized.

In Wiener filtering, it is assumed that the noise and image are uncorrelated, and that the intensity levels in the estimate are a linear function of the levels in the corrupted image.

To minimize the MSE, a convenient Wiener filter is given by:

$$\left[\frac{H^*(u,v)\,S_f(u,v)}{S_f(u,v)\left|H(u,v)\right|^2+S_n(u,v)}\right]. \tag{25}$$

Therefore:

$$Z(u,v)=\left[\frac{H^*(u,v)\,S_f(u,v)}{S_f(u,v)\left|H(u,v)\right|^2+S_n(u,v)}\right]G(u,v)=\left[\frac{H^*(u,v)}{\left|H(u,v)\right|^2+\dfrac{S_n(u,v)}{S_f(u,v)}}\right]G(u,v)\ , \tag{26}$$

which can be written as:

$$Z(u,v) = \left[\frac{1}{H(u,v)} \frac{|H(u,v)|^2}{|H(u,v)|^2 + \dfrac{S_n(u,v)}{S_f(u,v)}} \right] G(u,v) \tag{27}$$

where $H(u,v)$ is the degradation function, $H^*(u,v)$ is the complex conjugate of $H(u,v)$, $|H(u,v)|^2 = H^*(u,v)H(u,v)$, $S_n(u,v) = |N(u,v)|^2$ is the power spectrum of the noise and $S_f(u,v) = |F(u,v)|^2$ is the power spectrum of the undegraded image.

Note that, if the noise power spectrum is zero, the Wiener filter reduces to the direct inverse filter:

$$Z(u,v) = \frac{G(u,v)}{H(u,v)}. \tag{28}$$

Therefore,

$$Z(u,v) = F(u,v) + \frac{N(u,v)}{H(u,v)}, \tag{29}$$

In most cases, the power spectra of the noise and the uncorrupted image are unknown. For instance, in ultrasound imaging where the uncorrupted image is unavailable. Therefore, a scalar *K*, equal to the ratio of the average noise power over the average image power, is used to generate a constant array substituting the function $S_n(u,v) / S_f(u,v)$:

$$Z(u,v) = \left[\frac{1}{H(u,v)} \frac{|H(u,v)|^2}{|H(u,v)|^2 + K} \right] G(u,v). \tag{30}$$

To get a better result, the autocorrelation functions of the noise and the uncorrupted image, if they are known, are frequently used instead of the power spectrum of these functions. Since $|F(u,v)|^2 = FT\left[f(x,y) \circ f(x,y)\right]$; FT and $\circ$ being respectively the Fourier Transform and the autocorrelation function, we can obtain the autocorrelation function of the uncorrupted image by computing the inverse Fourier transform of the power spectrum. Similar comments hold for the autocorrelation of the noise [3].

3.3. Speckle Noise Reduction Filters

Since speckle noise is non-additively combined with the underlying image, researchers have developed various nonlinear filtering techniques which seek to reduce the effect of speckle noise while preserving the informative structure of the underlying image [4].

Some of the best known standard de-speckling filters are the methods of Lee, Kuan, Gamma and Frost. These filters use the second-order sample statistics within a minimum mean squared error estimation approach. They all have three essential criteria: the dimensions of moving window, the filter's parameter and the number of times the filter is applied to the image.

Denoising could address the multiplicative nature of the noise, or transform the multiplicative noise model into an additive noise model by transforming the image to the log domain and computing the logarithmic inversion after the filtering process.

3.3.1. Lee Filter

3.3.1.1. Locally Adaptive Lee Multiplicative Filter

The Lee multiplicative filter [5] is based on a multiplicative noise image model:

$$g(x,y) = f(x,y) \times n(x,y) \ , \tag{31}$$

where g denotes the observed image, f the original image and n the multiplicative speckle noise [6].

Based on an assumption that the noise is white with unity mean and uncorrelated with the image f, the multiplicative Lee filter gives the best mean-squared estimate of f at each pixel $g(x, y)$ by:

$$z(x,y) = \overline{f} + k_{(x,y)}\left[g(x,y) - \overline{f}\right], \tag{32}$$

where $z(x, y)$ is the filtered pixel, $\overline{f}$ is the mean of the original image pixels under the filtering window (i.e., the local mean) and $k_{(x,y)}$ is the gain factor given by:

$$k_{(x,y)} = \frac{\sigma_f^2}{\overline{f}^2 \sigma_n^2 + \sigma_f^2}, \tag{33}$$

where σ_f^2 is the variance of the original image pixels under the filtering window (local variance), and σ_n^2 is the noise variance.

The local adaptation of the filter is based on the calculation of the local statistics, $\overline{f}$ and σ_f^2 from the data sample estimates $\overline{g}$ and σ_g^2 determined over a local neighborhood window. This calculation requires the knowledge of the Coefficient of Variation (CoV) which is the ratio of the standard deviation to the mean in homogeneous areas. In our case (unity mean noise), this parameter is σ_n. If the original image is not available, which is the case in ultrasound and SAR images, σ_n can be determined from the type of image to be filtered and the known speckle characteristics of the Number of Looks or Equivalent Number of Looks (ENL). This calculated parameter effectively controls the amount of smoothing applied to the image by the filter.

The parameter $k_{(x,y)}$ allows the filter to both smooth noise and preserve edges; near edges, $k_{(x,y)} = 1$ (σ_f^2 is high) and smoothing is not performed because the filter shuts itself off and the estimate $z(x, y)$ is approximately equal to the observed pixel $g(x, y)$. In homogeneous areas, σ_f^2 is low and smoothing is performed.

Note that, if the noise is not white with unity mean and uncorrelated with the image *f*, a simpler Lee filter [7] uses the mean of the observed image instead of the mean of the original image and the multiplicative model is first approximated by a linear model then the mean-squared estimate is applied to the linear model:

$$z(x, y) = g(x, y)W + \bar{g}(1 - W) , \tag{34}$$

where $g(x, y)$ is the center pixel of the filtering window, $\bar{g}$ is the mean value within the window, W is a weighting function equal to $1 - C_u^2 / C_i^2$, C_i is the image variation coefficient equals $\sigma_g / \bar{g}$, C_u is the estimated noise variation coefficient equal to $\sqrt{1/NLOOK}$, σ_g is the standard deviation within the window and NLOOK is the Number of Looks of the corrupted image used to estimate the noise variance.

It is important to mention that the filter size and the number of iterations will greatly affect the quality of processed images [8, 9].

3.3.1.2. Enhanced Lee Multiplicative Filter

To improve the performance in edge areas, the neighborhood used in high variance areas takes into account the orientation of a possible edge by using a threshold *T*; for each pixel with $\sigma_g^2 > T$, oriented gradients are computed to select a subset of the neighborhood pixels. Therefore, the smoothing is performed over the new smaller window allowing more precise filtering scheme [10].

3.3.1.3. Modified Lee Filter (Ratio Based Edge Detection)

To solve the problems listed above, a modified Lee filter uses the ratio-based edge detector which estimates the edge strength at any pixel of interest in the image by computing the four ratio edge strengths at each pixel (Maximum Strength Edge Pruned Ratio of Averages method (MSP-RoA)) [10]:

$$R_i = \min\left(\frac{P_i}{Q_i}, \frac{Q_i}{P_i}\right), \quad i = 1,2,3,4 \tag{35}$$

corresponding to four orientations; horizontal, vertical, diagonal (from center to edges) and diagonal (from edges to center). P_i and Q_i are the averages calculated over the sub-windows P and Q respectively. Then a vector (R, O) is calculated to characterize a possible edge at that pixel, where the component $R = \min(R_1, R_2, R_3, R_4)$ is the edge strength and O is the orientation which yields the minimum R_i value. A candidate edge pixel is classified as edge pixel if the magnitude $R \leq T_h$ where T_h is a chosen threshold ($0 \leq T_h \leq 1$), and if R is the minimum magnitude of all the pixels in a sub-window of (2D + 1) x 1 pixels perpendicular to the orientation O, where D is the parameter that controls the sub-window size. By using the edge map produced by the MSP-RoA edge detector, the definition of the local neighborhood over which local statistics are calculated is refined; in the filtering window, all the pixels are classified as belonging to the valid or non-valid region.

3.3.2. Kuan Filter

3.3.2.1. Standard Kuan Filtering

Similarly to the Lee filter, Kuan filter [11] is designed to smooth out speckle noise while retaining edges or shape features in the image by applying the MMSE criterion. It performs spatial filtering on each individual pixel in an image using the grey level values in a square window surrounding each pixel. It first transforms the multiplicative noise model into a signal-dependent additive noise model. Then the minimum mean square error criterion is applied to the model. The resulting grey-level value $g(x, y)$ for the smoothed pixel is:

$$z(x,y)=g(x,y)W+\bar{g}(1-W) \ , \tag{36}$$

where (36) is identical to (35) except that $W=(1-(C_u^2 / C_i^2))/(1+C_u^2)$.

NLOOK is a parameter used to estimate the noise variance of the corrupted image and it effectively controls the amount of smoothing applied to the image by varying C_u. Theoretically, the correct value for NLOOK should be the effective Number of Looks of the image. It should be close to the actual Number of Looks of the noisy image, but may be different if the image has undergone resampling. This parameter is manually adjusted, in order to control the strength of the smoothing applied to the image.

Note that the resulting filter has the same form as the Lee filter presented earlier, but using a different weighting function. However, Kuan filter is considered to be superior to Lee since it doesn't make any approximation to the original model. The only limitation with this filter is that NLOOK parameter (or ENL) is needed for computation [12]. A method that achieves an improvement of the efficiency of the iterated Kuan filter, based on differential equations, is proposed in [13].

3.3.2.2. Adaptive Estimation of NLOOK Parameter

A method for adaptive estimation of an NLOOK value that yields near-optimal performance compared to manual parameter selection is proposed in [14, 15, 16]. This method allows to eliminate the need for several runs of the Kuan filter to reach acceptable performance. It defines an adjusted NLOOK value as:

$$NLK\_AD=(1+A)ENL \ , \tag{37}$$

where A is an adjustment coefficient depending on the noise variance and image detail, given by:

$$A = A_d + A_v \ , \tag{38}$$

where A_d and A_v are the detail and the noise variance components in A, respectively.

A_d parameter depends on the percentage of edges in the underlying image and it varies in a logarithmic or second order polynomial fashion [17]. The former can be expressed as:

$$A_d = \alpha \ln(P_e) + \beta \ , \tag{39}$$

and the latter as:

$$A_d = \lambda P_e^{\,2} + \delta P_e + \mu \ , \tag{40}$$

where $\alpha = 0.4015$, $\beta = 0.054$, $\lambda = -0.01215$, $\delta = 0.20512$ and $\mu = -0.01394$ are determined by curve fitting using the least squares method.

Note that, as the image detail (Pe) increases, A_d increases and consequently, *NLK_AD* increases. Therefore, less smoothing is performed.

A_v parameter is modeled by a second order polynomial as:

$$A_v = a(v - 2100)^2 + b(v - 2100) \ , \tag{41}$$

where $a = -2\text{x}10^{-8}$ and $b = -42\text{x}10^{-6}$ are determined by curve fitting.

Note that a higher noise variance corresponds to a lower value of A since a lower A value leads to more smoothing [20].

3.3.3. Gamma MAP Filter

The Gamma MAP (Maximum A Posteriori) [18, 19] filter is based on a Bayesian analysis of the image statistics. It performs spatial filtering on each individual pixel in an image using the grey level values in a square window surrounding each pixel.

Similarly to the Kuan filter, NLOOK effectively controls the amount of smoothing applied to the image, but in Gamma filtering, this parameter affects two thresholds: The speckle coefficient of variation $Cu = \sqrt{1/NLOOK}$ and the upper speckle coefficient of variation $C_{\max} = \sqrt{1+2/NLOOK}$. A small NLOOK value leads to more smoothing and a larger NLOOK value preserves more image features.

If $C_i \leq C_u$ ($C_i = \sigma_g / \bar{g}$ is the image variation coefficient), the window is over a low standard deviation area and therefore, smoothing is applied: $z(x,y) = \bar{g}$ (i.e., mean of the window pixels). If $C_i \geq C_{\max}$, the window is over a high standard deviation area and therefore, edge-pixel is replicated: $z(x,y) = g(x,y)$ (i.e., center pixel value).

If $C_u < C_i < C_{\max}$, the filtered pixel value is given by:

$$z(x,y) = \frac{(W - NLOOK - 1)\bar{g} + \sqrt{D}}{2W}, \tag{42}$$

where $D = \bar{g}^2 (W - NLOOK - 1)^2 + 4W \cdot NLOOK \cdot \bar{g} \cdot g(x,y)$

and $W = (1 + C_u^2) / (C_i^2 - C_u^2)$.

A Refined Gamma Map filter (RGMAP) [21] first detects contours belonging to step edges and thin linear structures, then applies the Gamma MAP filtering process to local statistics extracted from rectangular masks that do not cross image contours.

A modified version of the RGMAP speckle filter is proposed in [22]. This Modified Refined Gamma MAP (MRGMAP) filter first exploits local operators belonging to the odd-symmetric filter category employed by the RGMAP to detect image segments, and then it computes local statistics over areas that are subsets of the image segments having any possible shape (not necessarily rectangular).

3.3.4. Frost Filters

3.3.4.1. Frost Filter

Frost filter is a local statistic filter which attempts to model the SAR imaging process. It tends to reduce the multiplicative speckle noise by adjusting its parameters according to local (quite small) area statistic about the target pixel.

When uniform regions are filtered, the Frost filter acts as a mean filter and when high contrast regions are filtered, the filter acts as a high-pass filter with rapid decay of elements away from the filter center. Thus, large uniform areas will tend to be smoothed out and speckle removed, while high contrast edges and other objects will retain their values without being smoothed [23].

The frost filter can be considered as an adaptive-weighted-mean filter since it uses an adaptive filtering algorithm which is an exponential damped convolution kernel that adapts itself to features by computing a set of weighting factors for each pixel within the filtering window surrounding each pixel:

$$M_n = \exp\left(-DAMP \cdot \left(\frac{\sigma_g}{\bar{g}} \right)^2 \cdot T \right) \tag{43}$$

where *DAMP* is a factor that determines the extent of the exponential damping for the image, σ_g is the standard deviation of the filter window, $\bar{g}$ is the mean value within the window and T is the absolute value of the pixel distance between the center pixel to its surrounding pixels in the filter window [23].

The value of the filtered pixel is replaced with a value calculated from weighted sum of each pixel value g_n and the weights of each pixel M_n in the filter window over the weighted sum of M_n:

$$z(x, y) = \frac{\sum g_n M_n}{\sum M_n} \tag{44}$$

In fact, M is an exponentially decaying function with T and it is adjusted according to the local variance in each area. It's obvious that if $\sigma_g / \overline{g}$ approaches 0 (low variance-uniform area), $M_n = 1$, and the filter acts as a mean filter and if $\sigma_g / \overline{g}$ is very large (high variance-high contrast area), little smoothing occurs and the value of the current pixel is retained [24].

3.3.4.2. Enhanced Frost Filter

The Enhanced frost filter is an extension of the Frost filter that further divides the SAR image into homogeneous, heterogeneous and isolated point target areas. It applies a different exponentially weighting factors M to optimally filter each:

$$M_n = \exp\left(-DAMP \frac{(C_i - C_u)}{(C_{\max} - C_i)} T \right), \tag{45}$$

where $C_i = \sigma_g / \overline{g}$ is the local coefficient of variation of the filter window, $Cu = \sqrt{1 / NLOOK}$ is the speckle coefficient of variation of the image using equivalent number of looks, and $C_{\max} = \sqrt{1 + 2 / NLOOK}$ is the image upper speckle coefficient of variation [7].

The filter divides the image into three classes:

- If $C_i \leq C_u$, the filtered pixel is the mean of the filtering window: $z(x,y) = \overline{g}$.
- If $C_i \geq C_{\max}$, edge-pixel is replicated: $z(x,y) = g(x,y)$ (i.e., center pixel value).
- If $C_u < C_i < C_{\max}$, the filtered is given by the weighted value (cf. Eq. (44)).

Note that the Enhanced Frost filter better preserves the edges and texture of the image, compared to the standard Frost filter [25].

3.4. Wavelet Denoising

Wavelet transform consists of a set of basic functions used to analyze signals in both spatial and frequency domains simultaneously. These functions isolate signal discontinuities since high pass filtering is applied in order to obtain detail information and low pass filtering is applied to retrieve a smoother approximation of the signal, making it possible to analyze the signal at different scales [25, 26].

Multi-resolution Discrete Wavelet Transform (DWT) consists in passing the image through a filter bank; the image is decomposed to detail coefficients (output of a high pass filter) and approximation coefficients (output of a low pass filter). Then, the filter outputs are down-sampled by 2 in the purpose of discarding half the samples. The decomposition is repeated to further increase the frequency resolution [26, 25, 28].

Considering an image corrupted by an additive noise (cf. Eq. (11)), wavelet-based denoising (Figure 3) consists in:

- Applying the DWT to the noisy image g.
- Thresholding the detail coefficients (Wavelet shrinkage).
- Inverse transforming (IDWT) the result to obtain an estimation z of the original image f.

When dealing with images corrupted by multiplicative noise, a logarithmic transformation is applied to the noisy image before wavelet decomposition, to transform the multiplicative noise, into additive noise and after wavelet reconstruction, an exponential transformation is applied to reverse the logarithmic operation.

In wavelet domain, the output of the low pass filter consists of the high magnitude and low frequency components (approximation coefficients) and the output of the high pass filter consists of the low magnitude and high frequency components (detail coefficients). Figure 4 shows a one level, 2D wavelet decomposition scheme [29] where L(.) represents the low pass filtering operator, and the subscript (L) represents a low pass filtered output. Similarly, H(.) and the subscript (H) represent high pass filtering and a high-pass filtered output, respectively. The symbol (↓2) represents a down-sampling operator by a factor of 2. As a result of filtering and down sampling (Figure 4), four sub-bands are obtained: scaling (approximation) coefficients (g_{LL}), horizontal detail coefficients (g_{HL}), vertical detail coefficients (g_{LH}) and diagonal detail coefficients (g_{HH}). The (HL), (LH) and (HH) sub-bands, represent the high frequency (and low magnitude) components and the (LL) sub-band represents the low frequency (and high magnitude) components. At the next level of decomposition, only the (LL) component is passed to the decomposition process.

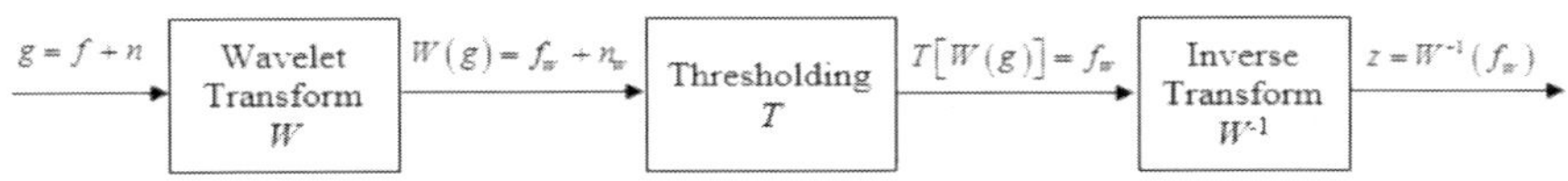

Figure 3. DWT denoising block diagram.

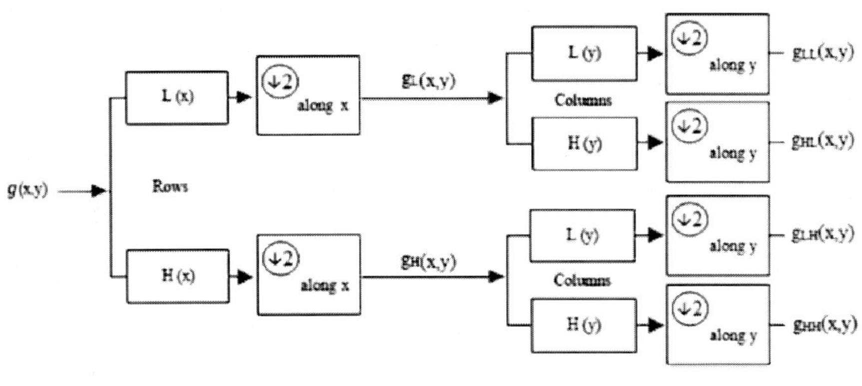

Figure 4. One level, 2D wavelet decomposition block diagram.

The reconstruction process or the Inverse Discrete Wavelet Transform (IDWT) consists of assembling back the wavelet coefficients to the original image. After each inverse low and high pass filtering, an up-sampling process (zeros insertion) is required to reverse the decomposition process.

Wavelet denoising consists in thresholding only the high frequency components while preserving most of the features in the image by retaining the approximation sub-band.

There are two thresholding methods frequently used; hard [27, 30] and soft [31] thresholding. In hard thresholding, the input is kept if its amplitude is greater than a threshold, otherwise it is forced to zero. In soft thresholding, if the absolute value of the input is less than or equal to a threshold, then the output is forced to zero, otherwise, the output is a scaled version of the input.

The choice of a suitable threshold is very important because a very large threshold will shrink to zero almost all coefficients, which results in over smoothing the image, while choosing a very small threshold will yield a noisy result which is close to the input [28]. VisuShrink [26] threshold selection uses a Universal (global) threshold applied to all sub-bands and scales after decomposition [32]. However, this threshold can be unwarrantedly large because it depends on the number of pixels yielding overly smoothed images [33]. BayesShrink [26, 30] is a sub-band adaptive threshold selection technique that determines a specific threshold for each sub-band [34]. It consists of finding, for each sub-band, a threshold that minimizes the expected value of the mean square error [31, 35].

3.5. Hybrid Wavelet-Spatial Filtering

In wavelet denoising, only detail coefficients are thresholded and approximation coefficients remain unchanged. This is intuitive since the approximation component is a low-frequency image, usually assumed to be noise-free. In fact, this is only a theoretical aspect since practically, approximation coefficients also contain speckle noise. The wavelet-spatial denoising filter proposed in [16] consists of smoothing the approximation coefficients in addition to detail coefficients as follows:

- Applying the Discrete Wavelet Transform to the speckled image
- Thresholding the detail coefficients (Wavelet shrinkage)
- Spatial Smoothing of the approximation sub-band coefficients
- Applying the Inverse Discrete Wavelet Transform to the result.

An important issue in the approximation image spatial smoothing is the preservation of the image characteristics. In other words, the algorithm must efficiently smooth out the noise in the approximation image without blurring its details or reducing the contrast and quality. Otherwise, the image structure and dynamics would be lost. The approximation sub-band coefficients in the wavelet domain can be smoothed using the adaptive Kuan filter presented earlier (cf. section 3.3.2.2) [16].

An acceleration of the wavelet filtering process used in this method was also proposed [16]. The idea is to find the number of decomposition levels that yields near optimal performance, without the need for several runs of the wavelet filter with different number of decomposition levels, regardless of the image size, its edge percentage, the noise variance, the wavelet type and base, the wavelet threshold selection technique and the wavelet thresholding technique. This acceleration algorithm consists in applying a level of wavelet decomposition and denoising, computing the relative difference between the corresponding ENL and the ENL of the previous decomposition level, and stop when the relative difference becomes less than 0.2 as shown in Figure 5. Note that the threshold of 0.2 was determined by an exhaustive study of the relative difference between two consecutive levels yielding a steady Peak Signal-to-Noise Ratio (PSNR) for different images, different image sizes, different noise variances, different wavelet types and bases.

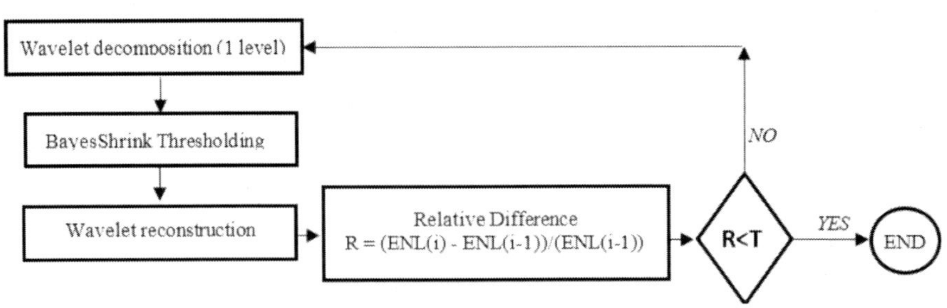

Figure 5. Flowchart of the acceleration algorithm.

3.6. Block Matching and 3D Filtering (BM3D) Denoising Method

The BM3D algorithm is a non-local denoising technique [11, 13] in a non-spatial domain. It combines non-local modeling and the sparse representation of the wavelet domain. The algorithm is divided into three processing stages: Block matching, 3D filtering, and aggregation.

In the block matching stage, the noisy image g is divided into blocks $G_{x \in X}$ where x represents the position of each block in the whole image X. For each block, the patches with most resemblance are grouped in a 3D array. Therefore, groups whose elements have a high degree of similarity are constructed separately. Two patches are similar if the Euclidean distance between them is less than a given threshold [13, 36]. The block matching advantages are the induction of a high correlation in the third dimension of the 3D array and the improvement of the dispersion of all possible configurations of details present in the image.

3D filtering is a procedure that jointly filters a group of similar blocks by exploiting the similarities between the grouped images and inside each image in the group. It consists of three different sections: 3D transform, shrinkage and inverse 3D transform. Those three procedures are jointly applied on the block [11]. The 3D transform consists of applying a 2D wavelet transform on each patch of the block then a 1D wavelet or Discrete Cosine Transform (DCT) on the resulting patches. As a result of the sparse representation given by the 3D transform, the shrinkage process can effectively attenuate the noise by eliminating the coefficients relying under a certain chosen threshold. The inverse 3D transform consists of assembling back the thresholded coefficients to reconstruct the 3D block.

Finally, the aggregation stage consists of combining the patches in the 3D filtered groups including the reference patch. A trivial solution is to compute the weighted mean of all the estimated patches overlapping at a pixel position.

4. Results

4.1. Evaluation Metrics

After filtering an image, it is necessary to know if the filtered image has been enhanced, or the blurring effect is dominating the noise reduction. Therefore, to study the quality-related outcomes of the filtering, we use four different metrics: the Peak Signal-to-Noise Ratio (PSNR), the Signal-to-Mean-Square-Error (S/MSE), the Coefficient of Correlation (CoC) and the Equivalent Number of Looks (ENL).

4.1.1. Peak Signal-to-Noise Ratio (PSNR)

PSNR is an engineering term for the ratio between the maximum possible power of a signal and the power of corrupting noise that affects the fidelity of its representation [8].

In image processing, the signal is the original image, and the noise is the error (blur) introduced by the denoising procedure. It is most easily defined via the mean squared error (MSE) which is defined for two $m\times n$ monochrome images I and K as:

$$MSE = \frac{1}{mn}\sum_{i=0}^{m-1}\sum_{j=0}^{n-1}[I(i,j) - K(i,j)]^2 , \tag{46}$$

The PSNR is defined as:

$$PSNR = 10\log_{10}\left(\frac{MAX_I^2}{MSE}\right) = 20\log_{10}\left(\frac{MAX_I}{\sqrt{MSE}}\right), \tag{47}$$

where MAX_I is the maximum possible pixel value of the image, which can be expressed as $2^k - 1$, where k is the bit depth of the image.

4.1.2. Signal to Mean Squared Error (S/MSE)

In order to quantify the achieved performance improvement, the Signal to Mean Squared Error can also be computed, based on the original and the noisy/denoised images:

$$S/MSE = 10\log_{10}\left(\frac{\sum_{i=0}^{m-1}\sum_{j=0}^{n-1}[I(i,j)]^2}{\sum_{i=0}^{m-1}\sum_{j=0}^{n-1}[K(i,j)-I(i,j)]^2}\right), \tag{48}$$

where I is the original image and K the noisy or denoised image [35].

4.1.3. Coefficient of Correlation (CoC)

The Coefficient of Correlation (CoC) metric is a qualitative measure for edge preservation computed as:

$$CoC = \frac{\Gamma\left(\Delta_I - \overline{\Delta_I}, \Delta_K - \overline{\Delta_K}\right)}{\sqrt{\Gamma\left(\Delta_I - \overline{\Delta_I}, \Delta_I - \overline{\Delta_I}\right)\cdot\Gamma\left(\Delta_K - \overline{\Delta_K}, \Delta_K - \overline{\Delta_K}\right)}}, \tag{49}$$

where Δ_I and Δ_K are the high-pass filtered versions of I and K respectively, obtained with a 3×3 pixel standard approximation of the Laplacian operator [37], $\overline{\Delta_I}$ is the mean value of Δ_I and $\overline{\Delta_K}$ is the mean value of Δ_K and the operator Γ is given by:

$$\Gamma(I,K) = \sum_{i=0}^{m-1}\sum_{j=0}^{n-1} I(i,j).K(i.j), \tag{50}$$

where $m \times n$ is the size of the corresponding images I and K.

The CoC cannot exceed 1 in absolute value. It is 1 in the case of an increasing linear relationship and -1 in the case of a decreasing linear relationship. Its value lies in between in all other cases, indicating the degree of linear dependence between the images. The absolute value of the correlation measure should be close to unity to an optimal effect of edge preservation [35].

4.1.4. Equivalent Number of Looks (ENL)

The Equivalent Number of Looks is also a good approach for evaluating the performance obtained after filtering. ENL is often calculated over a uniform region, but due to difficulties in identifying uniform areas in the image, we breakdown the image into small areas of 25 x 25 pixels, compute the ENL for each area and take the average of these ENL values. This latter method is the one used in all the text and implementation algorithms that follow. A large ENL value usually corresponds to a better performance. The formula for the ENL calculation is:

$$ENL = \left(\frac{\text{mean}}{\text{standard deviation}} \right)^2 \tag{51}$$

Note that PSNR, SMSE, and CoC are quantitative metrics based on a pixel by pixel calculation and the original image is needed for computation, whereas ENL is a qualitative metric based on the mean and standard deviation, and it can be computed without the need of the original image [25].

4.2. Practical Results

In this section, a major subset of the reviewed denoising techniques is evaluated and compared subjectively and objectively, using the evaluation metrics of section 4.1.

4.2.1. Spatial Filters

This section presents results of spatial filtering in the case of different noise distributions. Figure 6 shows an example of results obtained with the arithmetic and geometric means filters. The figure shows, from left to right, the Lena image, the same image but corrupted by an additive Gaussian noise of zero mean and a variance of 400, the result of filtering the noisy image with an arithmetic mean filter of size 3 × 3, and a geometric mean filter of the same size. The four metrics mentioned earlier were computed to compare the performance of the two filters. The PSNR of the noisy image is 22.1441, it increases to 29.5600 in the arithmetic mean filtered image and 28.7155 in the geometric mean filtered image. Similarly, S/MSE is 16.4878 in the noisy image, it increases to 23.9036 in the arithmetic mean filtered image and 23.0591 in the geometric mean filtered image. The CoC of the noisy image is 0.2472, it increases to 0.4598 and 0.4542 in the arithmetic and geometric mean filtered images respectively. ENL of the noisy image is 25.8483, it increases to 123.6523 in the arithmetic mean filtered image and 117.8511 in the geometric mean filtered image.

Table 1 presents denoising results obtained with arithmetic mean, geometric mean, harmonic mean, median, midpoint, alpha-trimmed, adaptive local noise reduction and adaptive median filters on a 512×512 Lena image corrupted by additive noise of variance 400 with different distributions.

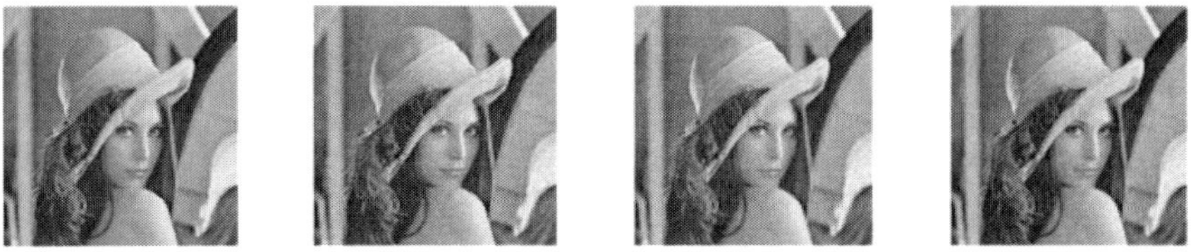

Figure 6. Spatial filtering results. From left to right: Lena image, same image corrupted by additive Gaussian noise, results of filtering with a 3 × 3 arithmetic and geometric mean filters.

Table 1. Denoising results on Lena image with additive noise of variance 400 and different distributions

		PSNR	S/MSE	CoC	ENL
Gaussian noise	Noisy image	22.1441	16.4878	0.2472	25.8483
	Arithmetic mean	29.5600	23.9036	0.4598	123.6523
	Geometric mean	28.7155	23.0591	0.4542	117.8511
	Harmonic mean	24.7497	19.0933	0.0020	110.5509
	Median	28.4118	22.7554	0.3823	94.0298
	Midpoint	27.1851	21.5287	0.2637	85.3018
	Alpha-trimmed	29.3392	23.6829	0.4683	113.5542
	Adaptive local	25.7821	20.1258	0.2436	53.1660
	Adaptive median	24.9157	19.2593	0.2597	45.8476
Uniform noise	Noisy image	22.1020	16.4457	0.2472	25.6527
	Arithmetic mean	29.5433	23.8869	0.4583	123.7151
	Geometric mean	28.9506	23.2942	0.4559	119.2686
	Harmonic mean	27.2487	21.5923	0.2544	116.0723
	Median	27.0119	21.3559	0.3324	69.4358
	Midpoint	29.3453	23.6890	0.3130	179.5109
	Alpha-trimmed	28.3093	22.6530	0.4302	90.2062
	Adaptive local	27.2161	21.5597	0.3225	71.3905
	Adaptive median	23.8230	18.1667	0.2160	37.8267
Rayleigh noise	Noisy image	22.1336	16.4772	0.2480	26.0571
	Arithmetic mean	29.5272	23.8709	0.4569	125.4416
	Geometric mean	29.2102	23.5538	0.4637	125.9020
	Harmonic mean	27.7854	22.1291	0.2494	127.3419
	Median	28.1703	22.5140	0.3771	91.9098
	Midpoint	26.9371	21.2808	0.2661	94.3996
	Alpha-trimmed	29.1373	23.4810	0.4605	110.8059
	Adaptive local	25.6596	20.0032	0.2343	51.8141
	Adaptive median	24.7938	19.1375	0.2553	44.5565
Exponential noise	Noisy image	22.2645	16.6082	0.2459	27.2122
	Arithmetic mean	29.5745	23.9182	0.4606	126.4666
	Geometric mean	29.7987	24.1423	0.4877	137.4639
	Harmonic mean	29.0576	23.4012	0.4825	144.4884
	Median	28.6015	22.9451	0.4282	114.6388
	Midpoint	23.8367	18.1803	0.2145	61.2771
	Alpha-trimmed	29.5158	23.8595	0.5039	134.6700
	Adaptive local	22.6265	16.9702	0.1186	28.5357
	Adaptive median	25.5574	19.9010	0.2875	50.9236

Erlang noise	Noisy image	22.2654	16.6091	0.2469	27.2739
	Arithmetic mean	29.5516	23.8952	0.4599	124.1077
	Geometric mean	29.7736	24.1172	0.4874	135.4064
	Harmonic mean	29.0627	23.4664	0.4834	148.7279
	Median	28.5924	22.9360	0.4290	114.0903
	Midpoint	23.8670	18.2107	0.2151	61.9881
	Alpha-trimmed	29.4558	23.8194	0.5011	130.7661
	Adaptive local	28.6257	16.9694	0.1208	28.0938
	Adaptive median	25.5868	19.9305	0.2912	51.5686

It can be seen in Table 1 that the impact of the additive noise with a variance of 400 on the image is almost the same regardless of the noise distribution. This can be noticed from the calculated metrics (for example, PSNR of the noisy images is 22.1441, 22.1020, 22.1336, 22.2625, and 22.2646 for the Gaussian, Uniform, Rayleigh, Exponential, Erlang distributions respectively).

Table 2. Denoising results on Lena image with speckle noise of variance 400 and different distributions

		PSNR	S/MSE	CoC	ENL
Gaussian noise	Noisy image	27.7732	22.1168	0.4384	55.3764
	Arithmetic mean	32.1354	26.4790	0.5706	226.3534
	Geometric mean	31.8512	26.1948	0.5710	232.2474
	Harmonic mean	31.1905	25.5342	0.5556	227.4139
	Median	32.0850	26.4286	0.5588	186.0035
	Midpoint	29.4344	23.7780	0.3151	165.4753
	Alpha-trimmed	32.5351	26.8787	0.6148	215.0795
	Adaptive local	12.0391	6.3827	0.1183	4.6893
	Adaptive median	29.9916	24.3353	0.4648	94.0698
Uniform noise	Noisy image	27.7581	22.1017	0.4390	55.3603
	Arithmetic mean	32.1518	26.4955	0.5714	230.4818
	Geometric mean	31.8632	26.2069	0.5722	234.3300
	Harmonic mean	31.2316	25.5752	0.5561	227.2646
	Median	31.2465	25.5902	0.5252	144.4847
	Midpoint	30.2108	24.5544	0.3399	290.3367
	Alpha-trimmed	32.0601	26.4037	0.5972	178.2673
	Adaptive local	12.1301	6.4737	0.1137	4.8410
	Adaptive median	29.0972	23.4408	0.4203	80.4180

Table 2. (Continued)

Rayleigh noise	Noisy image	27.7795	22.1231	0.4396	55.4523
	Arithmetic mean	32.1563	26.5000	0.5717	235.0085
	Geometric mean	31.9253	26.2689	0.5742	233.5555
	Harmonic mean	31.3259	25.6696	0.5601	241.4553
	Median	31.9447	26.2883	0.5568	176.8742
	Midpoint	29.4278	23.7715	0.3179	181.0586
	Alpha-trimmed	32.4593	26.8030	0.6142	211.5762
	Adaptive local	11.9448	6.2885	0.1221	4.5654
	Adaptive median	29.9462	24.2898	0.4609	92.6069
Exponential noise	Noisy image	27.9376	22.2812	0.4364	56.9059
	Arithmetic mean	32.1906	26.5343	0.5738	243.0346
	Geometric mean	32.0519	26.3955	0.5806	246.3449
	Harmonic mean	31.4672	25.8108	0.5689	255.9530
	Median	32.2352	26.5789	0.5939	226.2435
	Midpoint	27.9817	22.3253	0.2871	124.2568
	Alpha-trimmed	32.6485	26.9921	0.6358	253.2697
	Adaptive local	11.4828	5.8264	0.1374	4.0961
	Adaptive median	30.7084	25.0520	0.5027	106.3857
Erlang noise	Noisy image	27.9178	22.2614	0.4399	56.3230
	Arithmetic mean	32.1934	26.5370	0.5725	236.0208
	Geometric mean	32.0407	26.3844	0.5799	260.0776
	Harmonic mean	31.4697	25.8134	0.5674	253.7297
	Median	32.2411	26.5847	0.5915	225.3169
	Midpoint	28.0230	22.3667	0.2902	125.6862
	Alpha-trimmed	32.6696	27.0133	0.6371	251.5109
	Adaptive local	11.4788	5.8224	0.1381	4.0848
	Adaptive median	30.7197	25.0634	0.5039	107.3804

Both the arithmetic and geometric mean filters did a reasonable job of noise reduction. However, the performance of the filters depends on the noise distributions; the arithmetic mean filter is better than the geometric mean filter for a Gaussian noise, for example, whereas the geometric mean filter is better than the arithmetic mean filter for an Erlang noise distribution. In general, the geometric mean filter did not blur the image as much as the arithmetic mean filter. This is verified by the CoC metric which is greater in the geometric mean filtered images than the arithmetic mean-filtered images (with a Rayleigh, Exponential and Erlang noise distributions). The harmonic mean filter did a reasonable job of noise reduction with Exponential and Erlang noise distributions, whereas it gives worst results with Gaussian distribution.

The median filter provides the capability of noise reduction with less blurring than the harmonic mean filter. It does not show better results than the arithmetic and geometric mean filters since the median filter is more appropriate for salt and pepper noise.

The midpoint filter works best for the Gaussian and Uniform noise distributions as mentioned in section 3.1.2.3.

Although the alpha-trimmed filter is useful in situations involving a combination of multiple noise distributions, it gives acceptable results with most noise distributions (without noise combination), especially with Exponential noise.

The metrics obtained with the adaptive local noise reduction filter are less than those obtained with other filters, except for the Uniform noise.

Finally, the adaptive median filter does not outperform the static median filter since the noise is not of the salt-and-pepper type.

Table 2 presents denoising results obtained using the spatial filters on a speckled Lena image after applying the logarithmic transformation to transform the multiplicative noise into additive noise. In the purpose of comparing between the performance of these filters on multiplicative noise filtering and additive noise filtering, a noise variance of 400 is used but having a unity mean because the noise is multiplied with the uncorrupted image.

We observe that a noise variance of 400 is not a significant level with multiplicative noise (in the following sections, variances ranging from 2000 to 4500 are used with multiplicative noise). This can be seen from the high values of the different metrics computed for the noisy images. As a result, the computed metric values for the filtered images are higher than those obtained in the case of additive noise. However, in most filters and noise distributions, the rate of increase is greater in the case of additive noise than in multiplicative noise case, except for the ENL metric (for example, a PSNR increase of 7.4159 dB in additive noise arithmetic mean filtering and 4.3622 dB in multiplicative noise arithmetic mean filtering with Gaussian noise distribution). In fact, the ENL metric is not accurate and reliable in this case because it is affected by both the noise variance and the percentage of edge pixels in the image to be filtered. Furthermore, the adaptive local noise reduction filter fails to give acceptable performance when used in multiplicative noise reduction, regardless of the noise distribution.

Moreover, the impact of the 400-variance (multiplicative noise) on the image is almost the same regardless of the noise distribution, but the filtered images and the performance of the filters depend on the noise distributions similarly to the additive noise case: the alpha-trimmed and arithmetic mean filters seem to be the best choice dealing with Gaussian and Uniform noise respectively.

4.2.2. Wiener Filter

In this section, the Wiener filter is applied to the Lena image corrupted by an additive Gaussian noise of zero mean and a variance of 400 and degraded using a blurring filter of 7 × 7 pixels. Figure 7 shows, from upper-left to bottom-right, the Lena image, the corrupted image, the result of Wiener filtering using $S_n(u,v)/S_f(u,v)=0$, using a constant ratio, and using autocorrelation functions [36].

Table 3. Evaluation metrics for different Wiener filtering methods

	PSNR	S/MSE	CoC	ENL
Degraded noisy image	18.9340	13.2777	0.1390	43.9381
Direct inverse filtered image	15.1370	9.4807	0.0745	24.6088
Wiener filtered image using constant ratio	17.9188	12.2624	0.1069	34.6569
Wiener filtered image using autocorrelation functions	29.2242	23.5678	0.5855	171.1119

It can be clearly seen that Wiener filtering using constant ratio gives a significant improvement over direct inverse filtering, and the result obtained using autocorrelation functions is close to the original image, although some noise is still evident. Because the original image and noise functions were known, we are able to estimate the correct parameters and the autocorrelation functions method leads to the best results. These results are verified in Table 3 which presents the calculated metrics for the images shown in Figure 7.

Figure 7. Wiener filtering results. From upper-left to bottom-right: Lena image, degraded image using additive Gaussian noise and a blurring filter, Wiener filtering results with direct inverse filtering, using a constant ratio, and using autocorrelation functions.

4.2.3. Speckle Noise Reduction Filters

This section presents results obtained with speckle noise reduction filters in the case of different noise distributions.

Figure 8 shows an example of results obtained with the Lee multiplicative filter on the Lena image corrupted by speckle noise for different noise distributions. From top to bottom: Gaussian, Uniform, Rayleigh, Erlang distributions (on the left) and the corresponding Lee filtered images (on the right). The speckle noise is of unity mean and a variance of 3090 for all the distributions tested and an 11 x 11 filter is used as in [4]. Note that in our simulation, it is assumed that the noise variance is known. Table 4 shows the evaluation metrics calculated on the images of Figure 8. A noise reduction in the filtered images is seen for all noise distributions (ENL increases from 13.1844 to 180.6328 with Gaussian noise and 12.9630 to 231.0089 with Uniform noise, for example) and the advantage of retaining edges and shape features is verified by the increase of the CoC in the Lee filtered images for all noise distributions. However, we notice that the speckle noise on edges and contours is retained.

Figure 8. Images corrupted by speckle noise and the corresponding Lee filtered images. From top to bottom: Gaussian, Uniform, Rayleigh, Erlang distributions (on the left) and the corresponding Lee filtered images (on the right).

Table 4. Evaluation metrics for the Lee filtered images of Figure 8

	PSNR	S/MSE	CoC	ENL
Gaussian noise				
Noisy image	19.1545	13.4982	0.1728	13.1844
Lee filtered	26.6546	20.9983	0.3048	180.6328
Uniform noise				
Noisy image	19.0419	13.3856	0.1736	12.9630
Lee filtered	26.8105	21.1541	0.3149	231.0089
Rayleigh noise				
Noisy image	19.2948	13.6385	0.1707	13.4545
Lee filtered	26.6698	21.0134	0.3038	206.5495
Erlang noise				
Noisy image	19.2660	13.6097	0.1731	13.3602
Lee filtered	26.6543	20.9979	0.3057	191.3560

The application of the modified Lee filter [27] on a speckled image is tested and compared with the original Lee multiplicative filter. The speckle noise has unity mean and a variance of 2000, and both modified and original filters have a window size of 11 x 11. MSP-RoA parameters used are: window size 11x11, threshold T_h = 0.75 and correlation parameter $D = 1$. Table 5 shows the resulting PSNR of the two filtering methods. PSNR results show that the noise is reduced using both filters especially with the modified filter which is highlighted by a greater PSNR value. The edge-enhanced modified Lee filter performs well, especially in high variance regions since it uses edge information, therefore, the quality of the results and the improvement of PSNR using the modified filter are attributed to the ability of the edge-enhanced modified Lee filter to smooth noise in high variance areas on either side of edges including finely defined edges, as was shown in [4].

Figure 9 shows an example of speckled Lena image filtered using Kuan method with NLOOK = ENL (middle) and NLOOK adaptively estimated using Eq. (38). Table 6 shows the corresponding calculated evaluation metrics. It can be clearly seen that the image with the adjusted NLOOK parameter is much sharper and its details are better preserved.

Figure 9. Images corrupted by speckle noise and the corresponding Kuan filtered images. From top to bottom: Gaussian, Uniform, Rayleigh distributions (left), the corresponding Kuan filtered images using NLOOK=ENL (middle) and the corresponding Kuan filtered images using Eq. (38) for NLOOK estimation.

Table 5. PSNR comparison of Lee multiplicative and edge-enhanced modified Lee filters

	Speckled image	Lee multiplicative	Modified Lee
PSNR	19.3744	24.5514	25.6756

Table 7 shows the results obtained using a Gamma MAP filter on a speckled Lena image with NLOOK = ENL as in [25]. Similarly to the Lee and Kuan filters, a noise reduction is noticed for all noise distributions (increase of all metrics) and the increase in the CoC metric verify the importance of this adaptive filter in retaining edges. However, residual speckle noise is remaining in the filtered images on edges and contours regardless of the noise distribution.

Table 6. Evaluation metrics for Kuan filtered images using NLOOK = ENL and NLOOK estimated adaptively using Eq. (38)

	PSNR	S/MSE	CoC	ENL
Gaussian noise				
Noisy image	19.8372	14.1808	0.1907	14.9007
Kuan (NLOOK = ENL)	23.3680	17.7117	0.1476	39.8671
Kuan (Adaptive NLOOK)	27.2041	21.5478	0.3283	218.4143
Uniform noise				
Noisy image	19.7623	14.1060	0.1900	14.6731
Kuan (NLOOK = ENL)	23.2916	17.6353	0.1473	38.9549
Kuan (Adaptive NLOOK)	27.3573	21.7009	0.3347	276.2464
Rayleigh noise				
Noisy image	20.0161	14.3598	0.1901	15.2937
Kuan (NLOOK = ENL)	23.5801	17.9237	0.1589	42.6024
Kuan (Adaptive NLOOK)	27.2302	21.5739	0.3325	238.5623

Table 7. Evaluation metrics for Gamma filtered images corrupted by speckle noise of different distributions

	PSNR	S/MSE	CoC	ENL
Erlang noise				
Noisy image	19.7720	14.1156	0.1852	14.6286
Gamma filtered image	28.4444	22.7880	0.4070	248.5083
Gaussian noise				
Noisy image	19.7221	14.0658	0.1866	14.5736
Gamma filtered image	28.4502	22.7939	0.4056	249.4234
Uniform noise				
Noisy image	19.5999	13.9435	0.1835	14.2852
Gamma filtered image	28.4549	22.7986	0.4062	243.0913
Rayleigh noise				
Noisy image	19.8521	14.1957	0.1845	14.8401
Gamma filtered image	28.5537	22.8973	0.4095	246.4748

Table 8. Evaluation metrics for Frost and Enhanced Frost filtered images

	PSNR	S/MSE	CoC	ENL
Gaussian noise				
Noisy image	19.7045	14.0489	0.1849	14.5360
Frost filtered image	26.4510	20.7946	0.4305	245.1726
Enhanced Frost filtered image	27.2298	21.5734	0.4609	250.5225

Table 9. Results of speckle noise filtering techniques on a 512 × 512 Lena image corrupted with speckle noise of variance 3000 with different distributions

		PSNR	S/MSE	CoC	ENL
Gaussian noise	Noisy image	19.2482	13.5918	0.1781	13.3615
	Lee multiplicative	26.9749	21.3185	0.3028	98.0283
	Kuan (NLOOK = ENL)	22.8675	17.2111	0.1155	31.6124
	Kuan (Log. approximation)	27.2248	21.5685	0.3106	106.6198
	Kuan (Poly. approximation)	27.2389	21.5826	0.3154	110.6581
	Gamma MAP	28.1836	22.5272	0.3892	233.4135
	Frost	26.3968	20.7404	0.4225	240.3297
	Enhanced Frost	27.0823	21.4259	0.4394	233.8940
Uniform noise	Noisy image	19.1606	13.5042	0.1760	13.1826
	Lee multiplicative	27.5563	21.9000	0.3348	127.8801
	Kuan (NLOOK = ENL)	23.0604	17.4040	0.1210	33.0601
	Kuan (Log. approximation)	27.7578	22.1015	0.3420	144.8634
	Kuan (Poly. approximation)	27.7968	22.1405	0.3403	145.0305
	Gamma MAP	28.2011	22.5448	0.3934	227.2049
	Frost	26.4070	20.7506	0.4214	235.5507
	Enhanced Frost	27.0527	21.3963	0.4454	231.7399
Rayleigh noise	Noisy image	19.4281	13.7718	0.1760	13.7621
	Lee multiplicative	27.1461	21.4898	0.3158	106.4877
	Kuan (NLOOK = ENL)	23.1365	17.4802	0.1256	34.0457
	Kuan (Log. approximation)	27.3290	21.6726	0.3200	117.1893
	Kuan (Poly. approximation)	27.3523	21.6960	0.3236	116.7191
	Gamma MAP	28.2304	22.5741	0.3979	230.0371

	Frost	26.3868	20.7305	0.4231	240.8726
	Enhanced Frost	27.0836	21.4272	0.4576	234.4497
Exponential noise	Noisy image	20.0551	14.3987	0.1746	15.3187
	Lee multiplicative	25.9639	20.3076	0.2597	67.2103
	Kuan (NLOOK = ENL)	23.2500	17.5936	0.1368	34.4915
	Kuan (Log. approximation)	26.2178	20.5614	0.2689	73.9484
	Kuan (Poly. approximation)	26.1044	20.4480	0.2684	69.5444
	Gamma MAP	28.2633	22.6070	0.3779	201.6422
	Frost	26.3943	20.7380	0.4299	238.9131
	Enhanced Frost	27.0939	21.4376	0.4584	239.3688
Erlang noise	Noisy image	19.7728	14.1164	0.1736	14.5775
	Lee multiplicative	26.5388	20.8824	0.2821	84.0722
	Kuan (NLOOK=ENL)	23.3558	17.6994	0.1346	35.1668
	Kuan (Log. approximation)	26.6939	21.0375	0.2910	86.7863
	Kuan (Poly. approximation)	26.6776	21.0213	0.2911	88.9332
	Gamma MAP	28.2714	22.6151	0.3884	221.2841
	Frost	26.4154	20.7590	0.4266	236.1633
	Enhanced Frost	27.1403	21.4840	0.4566	244.0958

Table 8 shows the results obtained using Frost and Enhanced Frost filters on a speckled the Lena image with an exponential damping factor of 1 as in [38]. It can be clearly seen that Frost filter succeeds in reducing the noise. In addition, the Enhanced Frost filter gives better performance in reducing speckle (PSNR, S/MSE and ENL values are greater compared to the Frost filter). As expected, the CoC metric of the Enhanced filtered image is greater than the CoC of the Frost filtered image proving the superiority of the Enhanced filter in preserving edges and texture of the image. However, as Lee, Kuan and Gamma filtered images, residual speckle noise is remaining in the Frost and Enhanced Frost filtered images on edges and contours.

In order to compare the performance of the speckle denoising filters, we apply the Lee multiplicative filter, Kuan filter with NLOOK = ENL, Kuan filter with the logarithmic approximation (cf. Eq. (39)). Kuan filter with the polynomial approximation (cf. Eq. (40)), Gamma MAP filter, Frost and enhanced Frost filters on the Lena image corrupted by speckle noise of unity mean and a variance of 3000, with Gaussian, Uniform, Rayleigh, Exponential and Gamma distributions using a 7 × 7 window size. A comparison of results is presented in Table 9.

The performance of the filters is affected by the noise distribution. In fact, Lee multiplicative filter is better than the Enhanced Frost filter with Uniform noise distribution and worse with Gaussian noise. However, in general, Kuan filter gives better results than Lee multiplicative filter, except for the case with NLOOK = ENL because this latter parameter is not adjusted in this case. According to the noise distribution, one of the logarithmic and polynomial approximations has better performance than the other in Kuan filtering. In addition, the most important noise reduction is done by the Gamma MAP filter (highest PSNR and S/MSE), whereas the Frost and enhanced Frost filters preserve image features more than other filters as can be observed in CoC values.

Finally, as expected, the enhanced Frost filter gives better results in reducing speckle and preserving image features compared to the standard Frost filter.

4.2.4. Spatial and Speckle Noise Reduction Filters

Table 10 presents a comparison between PSNR results given by additive noise filters (after log transformation) and speckle filters. Three additive noise filters (arithmetic mean, median and alpha-trimmed) and five speckle filters (Lee multiplicative, Kuan, Gamma, Frost and enhanced Frost) are used in this comparison. Lena image corrupted by speckle noise of unity mean and a variance of 3000, and a 7 × 7 window size are used.

As a result, Gamma MAP filter gives the best PSNR regardless of the noise distribution. In general, speckle filters have better performance than additive noise filters except with Exponential and Gamma noise distributions where some additive noise filters give better PSNR results than speckle filters.

Table 10. Comparison between PSNR results given by additive noise filters (after log transformation) and speckle filters, using the Lena image corrupted by speckle noise of variance 3000

PSNR	Gaussian noise	Uniform noise	Rayleigh noise	Exponential noise	Gamma noise
Noisy image	19.2482	19.1606	19.4281	20.0551	19.7728
Lee multiplicative	26.9749	27.5563	27.1461	25.9639	26.5388
Kuan (Log. approximation)	27.2248	27.7578	27.3290	26.2178	26.6939
Kuan (Poly. approximation)	27.2389	27.7968	27.3523	26.1044	26.6776
Gamma MAP	28.1836	28.2011	28.2304	28.2633	28.2714
Frost	26.3968	26.4070	26.3868	26.3943	26.4154
Enhanced Frost	27.0823	27.0527	27.0836	27.0939	27.1403
Arithmetic Mean	26.7904	26.8040	26.8376	26.8781	26.8460
Median	27.1825	26.4273	26.7356	25.7060	26.1647
Alpha Trimmed	26.9128	26.8766	26.8421	26.7457	26.7830

4.2.5. Wavelet Filter

This section starts by showing results of wavelet denoising on the Lena image corrupted by a Gaussian additive noise having zero mean and variance 400, using Haar, db4, sym4, and bior6.8 wavelets. Soft and Hard VisuShrink (Universal Threshold) and BayesShrink, with two levels of decomposition, are used.

The first row of Figure 10 shows the Lena image (on the left) and the same image corrupted by additive Gaussian noise of zero mean and variance 400 (on the right). The second to fourth rows respectively show the db4 wavelet denoised image using Universal Soft Thresholding, Universal Hard Thresholding and BayesShrink with one decomposition level (left column) and two decomposition levels (right column). It is remarkable that two decomposition levels give better performance than one decomposition level. It can also be noticed that Universal Hard thresholding gives clearer results than Universal Soft Thresholding, and BayesShrink gives more pleasant results than Universal Hard Thresholding.

Figure 10. 1st row: Initial image (left) and the image corrupted by Gaussian additive noise (right). 2nd to 4th row: db4 wavelet denoised image using UT (soft), UT (hard) and BayesShrink, with one decomposition level (left) and two decomposition levels (right).

Figure 11 shows, from upper-left to bottom-right, the Haar, db4, sym4 and bior6.8 wavelet denoised images using BayesShrink with two decomposition levels. It's remarkable that the four resulting images are almost identical. However db4 and sym4 give better results than Haar and bior6.8. In fact, the PSNR of the noisy image was 22.1322. It increases to

28.5098 with Haar wavelets and 29.7562 with bior6.8 wavelets, while it reaches 29.7924 and 29.7908 with db4 and sym4 respectively.

Figure 11. Upper-left to bottom-right: Haar, db4, sym4 and bior6.8 wavelet denoised images using BayesShrink with two decomposition levels.

In the purpose of implementing wavelet denoising on speckled images, a log transformation (followed by exponential at the end of the algorithm) is used. Lena image corrupted by a Gaussian speckle noise having unity mean and a variance of 3100, Haar, db4, sym4, and bior6.8 wavelets, Soft and Hard VisuShrink, BayesShrink, with two levels of decomposition are used. The first row of Figure 12 shows the Lena image (on the left) and the same image corrupted by Gaussian speckle noise of unity mean and variance 3100 (on the right). The second to fourth rows respectively show the db4 wavelet denoised image using Universal Soft Thresholding, Universal Hard Thresholding and BayesShrink with one decomposition level (left column) and two decomposition levels (right column). Similarly to the additive noise case, two decomposition levels give better performance than one decomposition level. However, Universal Soft and Hard thresholding give more pleasant results than BayesShrink (PSNR of the noisy image was 19.1729, it increases to 24.5256 with Universal Soft thresholding and 23.9916 with BayesShrink, using one decomposition level).

Figure 12. 1st row: Initial image (left) and the image corrupted by Gaussian speckle noise (right). 2nd to 4th row: db4 wavelet denoised image using UT (soft), UT (hard) and BayesShrink, with one decomposition level (left) and two decomposition levels (right).

Figure 13. Upper-left to bottom-right: Haar, db4, sym4 and bior6.8 wavelet denoised images using BayesShrink with two decomposition levels.

Figure 13 shows, from upper-left to bottom-right, the Haar, db4, sym4 and bior6.8 wavelet denoised images using BayesShrink with two decomposition levels. It's visually remarkable that sym4 and bior6.8 give better results than Haar and db4.

By applying the evaluation metrics on results obtained with different wavelet types, using VisuShrink and BayesShrink, soft and hard thresholding, different noise variances with one and two decomposition levels, we can conclude that sym4 and bior6.8 lead to better results than Haar and db4. In addition, Universal Soft thresholding and Universal Hard thresholding give better performance than BayesShrink. Moreover, regardless of the noise variance and the wavelet type, Universal Soft thresholding is better than Universal Hard thresholding when using one decomposition level. On the other hand, Universal Hard thresholding is the best when using two decomposition levels. Obviously, two decomposition levels gives better results than one decomposition level.

Table 11. PSNR for wavelet denoised images corrupted by Gaussian speckle noise of variance 1100

Image	Woman-DH				Living Room			
	Haar	Db4	Sym4	Bior6.8	Haar	Db4	Sym4	Bior6.8
Noisy	24.3609	24.3786	24.3887	24.3919	23.7067	23.7179	23.6896	23.7248
Universal soft (1L)	29.3561	29.5657	29.8556	29.8624	26.0070	27.0209	26.9657	27.1261
Universal hard (1L)	29.1249	29.4747	29.5332	29.5346	26.2029	27.1004	27.0522	27.1831
BayesShrink (1L)	27.8361	28.0290	28.1952	27.7261	27.0736	27.3891	27.4636	27.3275
Universal soft (2L)	31.0059	33.2157	33.2755	33.4404	25.0075	25.9570	25.9695	26.0917
Universal hard (2L)	30.7643	32.4377	32.5442	32.5063	25.8454	26.6666	26.6949	26.8434
BayesShrink (2L)	28.8425	29.2456	29.6216	28.8459	27.9177	28.3090	28.4117	28.1748
Universal soft (3L)	29.8655	32.5033	32.7530	32.8928	24.0092	24.6883	24.7918	24.9085
Universal hard (3L)	30.3578	32.5775	32.5739	32.6323	25.3718	26.0412	26.1421	26.3053
BayesShrink (3L)	29.1183	29.5234	29.8989	28.9308	28.0679	28.4500	28.5598	28.2993
Universal soft (4L)	28.9802	31.2827	31.3146	31.5790	24.2008	24.1190	24.2006	24.3200
Universal hard (4L)	30.1739	32.0729	32.1161	32.3048	25.1579	25.8009	25.9048	26.0786
BayesShrink (4L)	28.9953	29.5674	29.9759	29.0187	28.1260	28.5027	28.5983	28.2520

Table 11 presents the PSNR for Haar, db4, sym4 and bior6.8 wavelet denoised images, using VisuShrink (Soft and Hard thresholding) and BayesShrink for two different images (WomanDarkHair and LivingRoom) taken from [39], corrupted by Gaussian speckle noise having unity mean and a variance of 1100 for one, two, three and four decomposition levels.

It can be observed in Table 11 that more decomposition levels lead to more smoothing. Thus, after several levels of decomposition, the PSNR begins to decrease due to oversmoothing. This level which gives the best PSNR depends on the image to be denoised; it is lowest with images having lot of edges and fine structures (LivingRoom) compared to images having low percentage of edges (WomanDarkHair). In addition, this level also depends on the wavelet type and family. In fact, four Haar levels seem to be equivalent to two db4 levels, which is logical since Haar is a lower base daubechies order (db1). Moreover, sym4 and bior6.8 lead to better performance than Haar and db4. However, their performance depends on the threshold selection technique; sym4 appears better than bior6.8 with BayesShrink and worse with Universal Thresholding.In conclusion, there is no one optimal threshold selection technique that always gives the best denoising performance. In fact, the best performance depends on which wavelet thresholding algorithm is chosen, the wavelet family (Haar, dbN …), the noise variance, the number of decomposition levels and the image to be denoised.

4.2.6. Wavelet-Spatial Filter

Table 12 presents results of hybrid wavelet-spatial denoising applied on a 512 × 512 Lena image corrupted by Gaussian speckle noise of unity mean and variance 3100, using BayesShrink threshold selection and 'sym2' wavelet type. In this table, we compare the wavelet denoising (without spatial smoothing) to the wavelet-spatial denoising using the adaptive Kuan filter on 1 decomposition level, on 3 decomposition levels where smoothing is applied on the last level (3rd level of 64×64 size), and on 3 decomposition levels where smoothing is applied on all the reconstructed levels.

Table 12. Comparison of wavelet denoising and wavelet-spatial denoising

	PSNR	CoC	ENL
Noisy image	19.1163	0.1754	13.0852
1L without spatial smoothing	23.9578	0.2713	33.5392
1L with spatial smoothing	25.2180	0.2672	55.3935
3L without spatial smoothing	26.7719	0.2959	75.7356
3L with spatial smoothing on the last level	26.7719	0.2959	75.7356
3L with spatial smoothing on all reconstructed levels	26.9596	0.2898	83.0166

From these results, we can deduce that a Kuan smoothing on the approximation image with one decomposition level leads to an increase in PSNR from 19.1163 dB to 25.2180 dB, compared to 23.9578 dB without smoothing the approximation image. It is important to note that Kuan smoothing on the last decomposition level does not result in any enhancement. In fact, the smoothing strength of the adaptive Kuan filter depends on the NLOOK parameter determined dynamically, which is very high in the approximation image of the last level, due to repetitive smoothing (low-pass filtering) at every decomposition. Therefore, no filtering is performed in this case. Moreover, a Kuan smoothing filter on the approximation image of each level before reconstruction, leads to an increase of 0.1877 dB compared to the case where approximation smoothing is performed only at the last decomposition level, without blurring the denoised image, which is obviously seen from the CoC metric.

CONCLUSION

Noise is a common problem found in different imaging applications such as medical ultrasound and radar imaging. It is a random process having a probability density function, specific to the underlying application, tending to reduce the image resolution and its main informative details. Considered as a random variable, noise can have Gaussian, Rayleigh, Erlang, exponential, uniform, impulse and speckle distributions. It can be added to, or multiplied by the image pixel values.

In spite of the efficiency of the existing image denoising techniques, most of these methods fail in different difficult cases and applications. They can lead to artifacts and remove important image information. Analyzing image denoising techniques represents a reliable assessment leading to improve existing techniques and propose new algorithms.

This chapter reviewed common noise distributions, presented a detailed survey of image denoising approaches with focus on additive and multiplicative noise reduction methods in the spatial, frequency and wavelet domains, as well as their applications. A comparison of various methods is carried out and evaluated qualitatively and quantitatively on synthetic as well as real images corrupted by noise of different distributions.

Consequently, a universal "best" approach has yet to be found and the search for an efficient image denoising method is still a valid challenge.

Image denoising is a key tool in image processing. We hope that this chapter will help and encourage research efforts in this field.

REFERENCES

[1] Shanthi, I. and Valarmathi, M. L. (2011). Speckle Noise Suppression of SAR Image using Hybrid Order Statistics Filters. *International Journal of Advanced Engineering Sciences and Technologies*, 5 (2): 229-235.

[2] Isar, A., Moga, S. and Isar, D. (2009). A new denoising system for SONAR images. *ACM Journal on Image and Video Processing*.

[3] Gonzales, R., Woods, G. and Richard, E. (2008) *Digital Image Processing,* Third Edition, Pearson Prentice Hall: 311-509.

[4] Ju, C. and Moloney, C. (1997). An Edge-Enhanced Modified Lee filter for the Smoothing of SAR Image Speckle Noise. *IEEE Workshop on Nonlinear Signal Image Processing*, Mackinac Island, MI.

[5] Lee, J. S. and Jurkevich, I. (1989). Segmentation of SAR images. *IEEE Trans. on Geoscience and Remote Sensing,* 27 (6): 674-680.

[6] Lee, J. S. (1981). Speckle analysis and smoothing of synthetic aperture radar images. *Computer Graphics and Image Processing*, 17: 24-32.

[7] Patil, L. V., Rane, K. K., Salvi, A. P. and Shitole, S. (2011). Evaluation of Statistical Digital Image Filters. *European Journal of Scientific Research.*

[8] Goyal, H. (2010). Qualitative and Quantitative Evaluation of Image Denoising Techniques. *International Journal of Computer Applications*, 8 (14): 31-34.

[9] Smith, D. M. (1996). Speckle reduction and segmentation of Synthetic Aperture Radar images. *International Journal of Remote Sensing*, 17 (11): 2043-2057.

[10] Ganugapati, S. S. and Moloney, C. R. (1995). A ratio based edge detector for speckled images based on maximum strength edge pruning. *Proceedings of ICIP- 95*, Washington, D. C., 2: 165-168.

[11] Kuan, D. T., Alexander, A., Sawchuk, T. C. and Strand, P. (1985). Adaptive Noise Smoothing Filter for Images with Signal-Dependent Noise. *IEEE Trans. on pattern analysis and Machine Intel.*, 7 (2): 165-177.

[12] Santosh, D. H. H., Kumar, V. N. L., Ramesh, D. R. and Prasad, P. S. (2011). Efficient Techniques for Denoising of Speckle and Highly Corrupted Impulse Noise Images. *IEEE Trans., Electronics Computer Technology (ICECT),* 3: 253-257.

[13] Klaine, L., Vozel, B. and Chehdi, K. (2003). Adaptive filtering of multiplicative noise by a new differential method. In *Proceeding of the Sixth International Symposium on Signal Processing and Its Applications.*

[14] Akl, A. and Yaacoub, C. (2013). A Hybrid Wavelet-Spatial Denoising Filter. *18th International Conference on Digital Signal Processing* (DSP), Santorini-Greece.

[15] Akl, A. and Yaacoub, C. (2013). A Hybrid Filter for Image Despeckling with Wavelet-Based Denoising and Spatial Filtering. *3rd International Conference on Communications and Information Technology (ICCIT),* AUB-Beirut-Lebanon.

[16] Akl, A. and Yaacoub, C. (2015). Accelerated Joint Image Despeckling Algorithm in the Wavelet and Spatial Domains. *International Journal of Image Processing (IJIP),* 9 (3): 102-126.

[17] Akl, A., Tabbara, K. and Yaacoub, C. (2012). An Enhanced Kuan Filter for Suboptimal Speckle Reduction. *2nd International Conference on Advances in Computational Tools for Engineering Applications (ACTEA),* NDU-Lebanon.

[18] Kuan, D., Sawchuk, A., Strand, T. and Chavel, P. (1987). Adaptive restoration of images with speckle. *IEEE Trans., Acoustic Speech, Signal Processing,* 35 (3): 373-383.

[19] Lopes, A., Nezry, E., Touzi, R. and Laur, H. (1990). Maximum a posteriori speckle filtering and first order texture models in SAR images. In *Proceeding of Geoscience and Remote Sensing Symposium '90,* College Park, MD, 3: 2409-2412.

[20] Akl, A., Tabbara, K. and Yaacoub, C. (2012). Edge-Based Automatic Parameter Selection for Sub-Optimal Image Despeckling using Kuan *Filter. 8th International Conference on Signal Image Technology and Internet Based Systems (SITIS),* Naples-Italy.

[21] Tsukasa, H. and Champika, W. (1993). Speckle Filtering and Texture Analysis in SAR Images. *IEEE Trans. on Geoscience and Remote Sensing Symposium,* 3: 1423-1425.

[22] Andrea, B. and Flavio, P. (1995). A Refined Gamma MAP SAR Speckle Filter with Improved Geometrical Adaptivity. *IEEE Trans. on Geoscience and Remote Sensing*, 33 (5): 145-1257.

[23] Aybike, S. and Hasan, S. B. (2011). Techniques of speckle noise reduction in ultrasound images. In *Proceeding of the 19th Signal Processing and Communications Applications Conference*: 375-378.

[24] D., Yunhan, Milne, A. and Forster, B. (2000). A Review of SAR Speckle Filters: Texture Restoration and Preservation. *IEEE Trans. on Geoscience and Remote Sensing,* 2: 633-635.

[25] Sern, H. (2001). *Denoising of Noise Speckle in Radar Image.* B. E. Thesis, School of Information Technology and Electrical Engineering, The University of Queensland.

[26] Isa, M., Noviyanto, A. and Arymurthy, A. (2011). Optimal Selection of Wavelet Thresholding Algorithm for ECG Signal Denoising. *IEEE Trans. on Advanced Computer Science and Information System*: 365-370.

[27] Daubechies, I. (1994). Ten lectures on wavelets. The Conference Board of the Mathematical Sciences, *Series in Applied Mathematics*, 61: 194-202.

[28] Mallat, S. (1989). A theory for multiresolution signal decomposition: The wavelet representation. *IEEE Trans., Pattern Anal. Machine Intell.,* 11: 674–693.

[29] Grace, S., Yu, B. and Vetterli, M. (2000). Adaptive Wavelet Thresholding for Image Denoising and Compression. *IEEE Trans on Image Processing*, 9: 1532-1546.

[30] Donoho, D. (1995). De-Noising by Soft-Thresholding. *IEEE Trans. on Information Theory*, 41 (3): 613-627.

[31] Donoho, D. (1994). Spatial adaptation by wavelet shrinkage. *Biometrika,* 81: 425-455.

[32] Yan, Y. Baolong, G., Zhanlong, Y. and Xiang, Fu. (2008). Image Noise Removal viaWavelet Transform and r/K-PSO. *IEEE Trans., Fourth International Conference on Natural Computation*: 544-547.

[33] Guangmin, S. and Fudong, L. (2003). Image Denoising with Optimized Subband Threshold. In *Proceeding of the Fifth International Conference on Computational Intelligence and Multimedia Applications.*

[34] Pizurica, A. and Philips, W. (2006). Estimating the Probability of the Presence of a Signal of Interest in Multiresolution Single- and Multiband Image Denoising. *IEEE Trans. on Image Processing*, 15 (3).

[35] Achim, A., Bezerianos, A. and Tsakalides, P. (2001). Ultrasound Image Denoising Via Maximum A Posteriori Estimation Of Wavelet Coefficients. *Proceedings of the 23rd Annual International Conference of the IEEE,* 3: 2553-2556.

[36] Gonzales, R., Woods, C., Richard, E. and Stevens, L. (2004). *Digital Image Processing Using Matlab*, Pearson Prentice Hall: 147-279.

[37] Rangarajan, R., Venkataramanan, R. and Shah, S. (2002). *Image Denoising Using Wavelets.* Ph.D. thesis.

[38] Sheng, Y. and Xia, Z. G. (1996). A Comprehensive Evaluation of Filters for Radar Speckle Suppression. In *Proceeding of Geoscience and Remote Sensing Symposium.*

[39] ImageProcessingPlace.com, Image Databases.

In: Noise Reduction
Editor: Marc Farrow
ISBN: 978-1-53613-541-1

Chapter 2

NOISE IN MULTIVIEW VIDEOS

Olgierd Stankiewicz*
Multimedia Telecommunications and Microelectronics,
Poznan University of Technology, Poznań, Poland

ABSTRACT

The chapter begins with an introduction to multiple camera systems: their usage, structure, hardware and applications. Some examples of existing systems are being presented along with quantitative comparisons of parameters like: sensor resolution, technology, frame rate, shutter-related features, need and means for temporal synchronization.

Next, means for noise estimation in multiview systems are presented. This part covers references to the estimation of noise parameters (statistical distributions, noise independence etc.) and also examples of measurements with a discussion of the presented data and caveats.

The third part of the chapter presents considerations about how the knowledge about noise existing in multiview camera systems can augment video processing used in applications in the field. Among other mentioned applications, including light-field video acquisition and compression, multiview video plus depth (MVD) acquisition and compression and also

* Corresponding Author Email: ostank@multimedia.edu.pl.

stereoscopic depth estimation. The chapter focuses on two subjects: stereoscopic depth estimation and compression. On these examples, some ideas about improving existing algorithms are shown along with some practical results that can be attained. In the case of stereoscopic depth estimation, the relation between noise in the camera system and the matching technique is shown on the example of depth estimation technique based on Maximum A posteriori Probability (MAP) rule. In the case of compression, it is shown how noise reduction and extraction as a preprocessing step can be used in order to improve compression efficiency and increase the subjective quality of multiview video.

Keywords: noise reduction, noise extraction, multiple view, multiview, immersive media, free viewpoint navigation

1. Introduction

The new generation video systems are anticipated to go beyond the limitations of the classical video, not only with respect to resolution, color reproduction, and frame rate, but also with respect to immersiveness. The user is no longer intended to be an external viewer. He/she is expected to stand inside and possibly interact with the scene. The boundary between the real world and virtual becomes fuzzier every year and we are closer to breaking it than ever before. Not surprisingly, immersive media concepts are appealing to the users, service provides and to the research community.

Hitherto, there is no consensus about the technology that will provide a sufficient break-through and grant the user with full visual immersiveness. In particular, the situation varies depending on whether we are considering, presentation, content representation and compression, or acquisition.

For presentation, there are stereoscopic displays (passive or active), autostereoscopic displays (allowing experiencing the depth of scene without wearing 3-D glasses), and VR headsets etc., each of which still have its deficiencies.

For the content representation, the state-of-the-art research works include many approaches: renderings based on 3-D models (e.g., mesh based), 360 panoramic video, spare multiview systems with virtual view synthesis (e.g., based on depth + image-based rendering), dense multiview systems aka light-field (e.g., with use of plenoptic lenses) and many more.

Aside from the ultimate technology that will used be for the representation format of the scene, the content somehow has to be captured. And as far as dynamic natural scenes are considered, currently, the most practical solution to that task is *acquisition of the scene with the use of multiple cameras.*

Multiview video acquisition differs from classical monoscopic video capture. The obvious differences include a much higher cost and a much bigger volume of data. The latter translates to bandwidth if transmission is considered, even when the newest achievements of compression technology are used.

The less obvious differences are sometimes more cumbersome than the obvious ones. For example, in many applications that use multiview, it is required that the frames are captured at exactly the same time in all cameras. Such requirements enforce usage of some means of synchronization between the cameras. The synchronization signal may be distributed with the use of wire or radio connections, but that also increases the complexity of the required infrastructure. Another example is the necessity of calibration of camera arrangements. Without the knowledge about the exact positions of the cameras in the scene, it is impossible to compare and merge information from them into a single unified representation of the scene; e.g., without this, we cannot tell whether two cameras observe the same object from different perspectives, or whether they are observing two different (although similar) objects.

Yet another example is the necessity for careful calibration of the parameters of each camera. This includes the intrinsic optical system parameters (focal length, lens distortion) and color representation characteristics, e.g., simple ones like brightness or color temperature or sophisticated color mapping schemes. This is especially significant in applications, where during the further processing, the content captured by cameras is directly compared, e.g., during correspondence matching in the estimation of the depth.

All of those mentioned above are challenging problems which are investigated by the research community and by the industry. Based on the progress that has already been made during recent years, we can optimistically expect that the development of fully immersive video systems in the future will be possible.

There is, however, one phenomenon that is omitted too often when multiview video systems are considered: *the presence of noise.*

Of course noise is present in classical monoscopic video systems. We may ask: so what are the differences, specific to multiview video? This chapter considers the answer to this question. For now let us notice that in many applications that use multiview video, including light-field, free-viewpoint television (FTV), three-dimensional television (3DTV) [1], advanced algorithms are employed that simultaneously process video from multiple views. Such algorithms include stereoscopic depth estimation [2], virtual view synthesis [3], super-resolution [4], light-field rendering [5], and advanced refocusing [6]. Another process applied on video content is compression, which can also be adapted to multiview [42, 43].

All of these operations and algorithms are conditioned by the presence of noise in the input video content. In the formulation of such algorithms, typically, either the presence of noise is omitted [7] or the assumptions about the characteristics of the noise are made only implicitly [8]. Even when the presence of the noise in the video material is assumed explicitly, the most common being Gaussian noise [9, 10, 11], is considered, often without any experimental verification [12].

This is very unfortunate. In practical applications, the devised algorithms work with very various video sequences, acquired with different cameras. If the exact characteristics of the noise (e.g., spatial spectrum, temporal correlations, value distributions) are ignored, the performance of such algorithms may be degraded in an unpredictable manner. This also lowers the reproducibility of the research, which may be a limiting factor for the speed of further developments in the field.

Later in the chapter, it is shown how such algorithms can be enhanced with knowledge about noise, but first, let us consider some characteristics of exemplary video systems with multiple cameras. Based on them, we will analyze what noise can be expected in such systems.

2. Multiple Camera Systems Examples

Features of multicamera systems are highly defined by the specifics of the current multiview technology, which is not yet mature today. There are no comprehensive industrial solutions yet, and on the other hand, construction of an experimental multiview system is a very demanding task, regarding both technology and human efforts. Currently there is no universal system design that would satisfy all needs. Development of a practical multicamera system requires making many trade-offs, because some of the constraints are inter-connected and cannot be considered separately. The most important features of multicamera systems include: number of the cameras, arrangement and mobility of the cameras, type of cameras and their resolution, sensitivity, shutter type, time synchronization interface etc., some of which are described below.

The *number of the cameras* is perhaps the most distinguishing parameter of the whole multi- camera system. To some extent, the more cameras we have, the better, but typically there are some practical limitations: budget, size and throughput (of processing and transmission/storage).

Some parts of multiview video processing has to be done on-the-fly e.g., in realtime. Therefore, depending on the particular application, *processing throughput* may be an important issue. This may include synchronization of the streams, rectification, color-correction and depth map estimation. All of those are influenced by data throughput. If we consider uncompressed data streams from 9 Full-HD cameras, 25 frames per second, we approximately get 6 Gbit/s of required throughput. Obviously, it is outstanding for transmission, but it is also quite a challenge when processing is considered, even with very fast machines. For example, GDDR5X memory, used in modern Graphical Processing Units (GPUs), can provide memory bandwidth of about 10 Gbit/s, which is not even enough to read the data and write it back after processing! Therefore, usage of multi-computer clusters is often preferred over single workstations.

The *mobility* of a system is often considered, which is especially important for systems intended for outdoor usage. Mobility means the ability for the system to be easily rearranged in a different place and to be reconfigured very quickly. Depending on the camera arrangement and mounting, some wheels, caterpillars or rails may be required (see further examples of systems for the reference).

The *camera arrangement* used depends mostly on the scene that is capturing and the target application. Light-field systems, composed from multiple individual cameras are typically arranged in dense arrays. In the case of stationary sparse multiview systems, cameras are typically arranged over a line or over an arc. The linear setup offers good visual sensations for distant scenes and are easier to handle from a software-point of view (rectification and depth map algorithms are commonly investigated for linear setup). On the other hand, arc-based setups give better 3D depth-sensation in cases of short-range, indoor scenes. It also more resembles a human visual system. Depending on mobility, camera number and spacing, the use of a rig should be considered. Although a rig can be bulky and heavy, it offers much better stiffness and rigidity than the use of separate camera tripods. In the case of mobile systems, frequently arbitrary or moving settings are considered.

Finally, the most sensitive part of a multicamera system is the *type of camera* with a particular optical sensor. Implicitly, the choice of camera can have influence on all parameters of the whole system. The size of the camera's optical sensor influences minimal camera spacing and thus impacts possible camera arrangements and mounting.

Resolution of the camera (SD, Full-HD or nowadays even 4k) influences the required processing throughput. Pessimistically, this can enforce off-line processing with use of intermediate video compression for transmission, or optimistically, can allow real-time processing of even uncompressed data streams. Also, as mentioned before, certain applications require exact time-synchronization of frame capture, which can be supported by a given camera or not. Tools for attaining *time-synchronization* are synchronization input/output ports (genlock), time-stamp signatures [13], or control interfaces like LANC [14].

The quality of the acquired image is of course dependent not only on the resolution but also on the *sensitivity and dynamic range*. This becomes especially important when capturing scenes which are very dim (e.g., during the night) or have huge variations of intensity within the scene, e.g., sun spotlights versus shadow.

Another especially important factor in the context of multiple camera systems is the type of shutter, e.g., rolling shutter or *global shutter*. In the case of classical monoscopic video systems, a rolling shutter merely leads to the annoying skew (from top to bottom) of fast moving objects. In the case of multiview camera systems, the importance of the shutter is higher, because the content captured by the cameras is not only presented to the viewer but it is also used to reconstruct a 3D model of the scene. Therefore, in order to minimize the negative influence of motion in the scene (which is captured with all cameras from different angles of observation) it is highly beneficial to use cameras that have a global shutter. Unfortunately, many cameras, especially cheap ones, provide rolling shutters which make shooting fast-moving scenes troublesome.

Last, but definitely not least: *prices of the cameras* vary very much, and typically for more advanced cameras we have to pay more. Quality costs, and as such limits the number of cameras that one can afford.

As already mentioned, it is impossible to build a universal system that would satisfy all possible needs. Nonetheless, we can consider some successful systems built experimentally and currently used for work on the development of multiview technology. Of course, known multicamera systems are not limited to those examples. A survey of them has been gathered in Table 1.

Table 1. Examples of multicamera systems

	Nagoya University [15]	**Fraunhofer HHI [16, 17]**	**Poznan University of Technology Linear rig [18]**	**Poznan University of Technology Modular system [19]**	**Poznan University of Technology Sports Hall system [20]**
Year of production	2005-2007	2008	2009	2013	2017
Number of cameras	100	16	9	10	36
Arrangement	Arc/linear	Linear	Linear	Any	Along walls, arranged in pairs
Camera spacing	50 mm	65 mm	137.5 mm	Arbitray $\geq$ 50cm	Inside pair: 0.5 m Between pairs: 3 m
Mobility	Immobile	Immobile	Mobile rig	Mobile individual modules	Stationary rig mounted in sports hall
Camera type	JAI Pulnix TMC-1400CL	Hitachi 3-chip CCD progressive scan RGB camera HV-F31CL-S1	Canon XH-G1 CCD	Canon XH-G1 CCD	Go Pro Hero 4 Black with backpack for synchronization
Resolution	1392 × 1040	1024 × 748	1920 × 1080	1920 × 1080	1920 × 1080
Frame rate	29.4 FPS	16.67 FPS	24 FPS	24 FPS	25 - 120 FPS
Interface	CameraLink	CameraLink	SDI	SDI	Ethernet
Recording	Personal computers	Workstation cluster	Workstation cluster	Built-in	Built-in
Processing infrastucture	Personal computers	Workstation cluster	Workstation cluster	Workstation cluster	Workstation cluster

2.1. Nagoya University Multiview System

Figure 1. Photographs of Nagoya University camera system [15].
Thanks to courtesy of Prof. M. Tanimoto, Nagoya University, Japan.

The system (Figure 1) built at *Nagoya University* [15], Japan, is composed of 80 cameras which are fixed on the steel frame with a 50 mm interval. Depending on the given setup the cameras can be placed linearly or on an arc which converges at about 8 m from the camera array. The cameras that are used are JAI Pulnix TMC-1400CL. They enable the acquisition of SXGA-images with a frame rate of almost 30 FPS. The shooting is temporally synchronized up to about 1 μs. The system also consists of one host-server PC and 100 client personal computers. The interface between the camera and the computer is Camera-Link.

2.2. Heinrich Hertz Institute Multiview System

Figure 2. Photographs of Fraunhofer HHI Camera system [16, 17]. Thanks to courtesy of Dr. Ing. K. Müller, Fraunhofer-Institute for Telecommunications, Heinrich Hertz Institute, Germany.

The system developed at the Fraunhofer Institute for Telecommunications, *Heinrich Hertz Institute, HHI,* Germany [16, 17], (Figure 2), consists of 16 machine vision cameras arranged linearly on a rig. The cameras used are HVF31CL-S1, Hitachi 3-chip CCD progressive scan RGB with XGA resolution. These are equipped with high quality external lenses (Fujinon DF6HA-1B 6mm) with a quite narrow field of view: 44° horizontally and 33° vertically. The trigger period is 20 ms which results in a frame rate of 16.67 FPS. The data is captured in raw RGB format, using a CameraLink interface via a professional workstation cluster.

2.3. Poznan University of Technology Multiview Systems

The Chair of Multimedia Electronics and Telecommunications, *Poznan University of Technology,* Poland has developed 3 multiview systems so far. All of them are intended as a framework for works on 3D and Freeview navigation television [18, 19, 20].

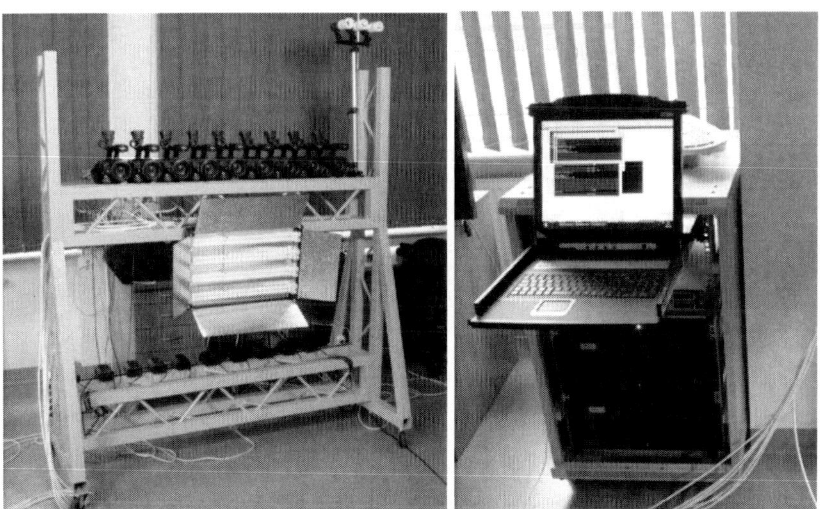

Figure 3. Experimental multiview camera system built at Poznan University of Technology, Poznań, Poland (left – camera rig, right – part of recording system). Photos by the author.

The first of those systems, [18] (Figure 3), consists of 9 cinematic Canon XH-G1 cameras placed on a mobile (wheeled) metal rig. The rig has been manufactured exclusively to provide special mounting pads that allow the precise alignment of the cameras. The output video signal is Full-HD (1920 × 1080) and is provided via SDI interface. All streams are temporally synchronized with the use of a genlock and captured by a raid-like workstation cluster. The whole processing is done off-line.

The second system [19] (Figure 4) comprises the same Canon XH-G1. Instead of a solid rig with a linear arrangement of cameras, independent camera rigs have been used and are equipped with dedicated hardware modules, allowing entirely mobile operation, e.g., synchronization, recovering, battery pack.

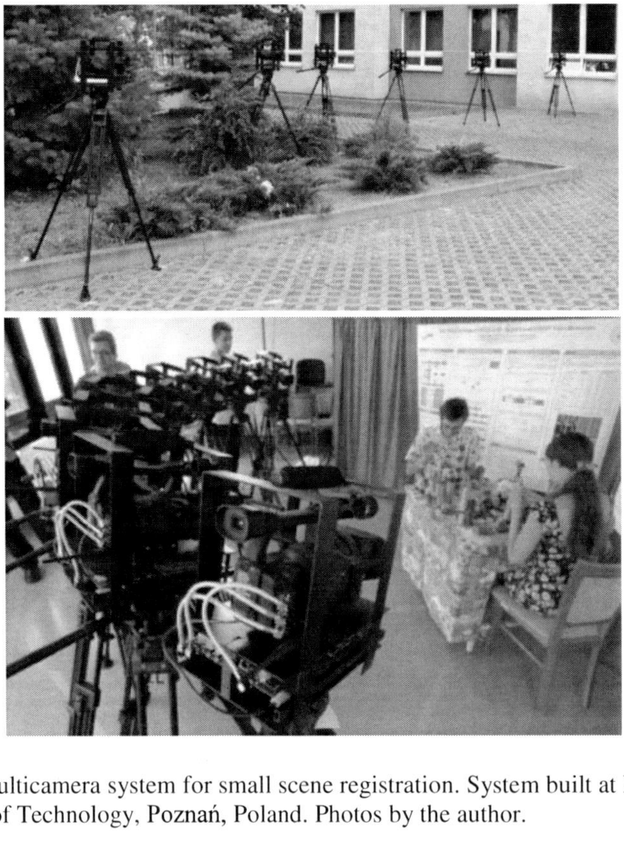

Figure 4. Multicamera system for small scene registration. System built at Poznan University of Technology, Poznań, Poland. Photos by the author.

Figure 5. Installation of the cameras in the sports hall of Poznan University of Technology, Poznań, Poland. Photos by the author.

The third, most recent system [20] has been developed in a sports hall (Figure 5) of the Poznan University of Technology. This multicamera system is mounted on a special rig installed at the height of circa 4.5 meter above the floor (Figures 2.4 and 2.5) with a 3 meter interval between camera pairs. As a camera module, consumer GoPro Hero4 Black cameras have been used. In order to support synchronization acquisition, a dedicated hardware backpack has been connected to each camera.

3. Analysis of Noise Characteristics

Relative to its importance, noise analysis and modelling itself has gained a relatively small interest among researchers related to video processing. Thus, the higher is the value of the work which considers it. For example, in work [21], noise introduced by the CMOS sensor is considered. Probability distribution of the noise in the form of Gaussian function is analyzed, but unfortunatelly no experimental verification results are provided. As another example, work [22] proposed to analyze noise by capturing a testing pattern, which comprised several patches with uniform colours. Such methodology can be applied only if the camera system is available for testing and calibration. In particular, it cannot be used in the case when the content is already acquired and provided without any other specific camera system knowledge. Such a scenario is quite common in practice. Consumer video content is rarely accompanied by information about the acquiring device. If no knowledge about the camera nor the noise characteristics is provided, the video processing algorithms cannot be enhanced and tuned.

But is there a way to attain the noise characteristics from such content? Yes! In this section a very simple, yet practical method will be presented. As it will be shown, is provides sufficient accuracy and allows us to draw interesting conclussions. The presented methodology allows for noise extraction and verification of the assumptions about the characteristics of noise in the multiview video material, without knowing any a priori information about the camera system.

The main idea of the presented methodology is very simple: the noise existing in a video sequence can be attained as a difference between the original and a denoised version of the given sequence. There are very many methods for noise reduction in video sequences, even apart from the ones presented in this book (e.g., [9, 23, 24, 25, 26, 27, 28, 29]). In principle, all of such methods can be used for noise extraction. Although very many advanced techniques exist, in our current considerations the purpose of noise reduction is in fact solely noise extraction for further analysis, and not noise extraction itself. Therefore, there is no need for the use of advanced techniques, and thus the simplest and the most straight-forward one is used. Of course, this simple approach can be extended and enhanced with more advanced techniques, like spatial filtering, temporal filtering, Kalman filtering, motion estimation and background extraction techniques etc.

Let us consider a very straight-forward technique [30, 31], in which only static fragments (without any movement) from the analyzed video are selected and used. In such fragments, previous and succesive frames of video represent the same image, altered only with different noise. Therefore, the image without noise can be easily retrieved, as an average of the frames.

We denote an image in time frame i as Y_i and $Y_i(x,y)$ as value (e.g., luminance) in coordinates x,y. Thus, for the average image $A(x,y)$, calculated from n previous and m following frames in regions considered to be static fragment, the calculation is as follows (1):

$$A(x,y) = \frac{1}{n+m+1}\sum_{k=i-n}^{i+m} Y_k(x,y). \tag{1}$$

The average image $A(x,y)$ can be interpreted as the expected value estimator $E[Y(x,y)]$ of many different realizations of the random process $Y_i(x,y)$. We search for the estimation of noise in the sequence, $N_i(x,y)$, which can be simply calculated as:

$$N_i(x,y) = Y_i(x,y) - A(x,y). \tag{2}$$

In a case, when noise $N_i(x,y)$ in independent in each other frame i, the estimator (2) is statistically unweighted. Otherwise, estimator (2) is statistically weighted, but it can be noticed that the error of the estimation decreases with the number of population samples: $n + m + 1$. Therefore, in practical use-cases, the error can be considerably reduced.

How can we use the attained noise values? We can measure the characteristics of the noise and use the gained knowledge *to enhance video processing!*

The first characteristic that we will be interested in, is the distribution of noise values. It can be calculated by taking a histogram of values $N_i(x,y)$. A histogram (denoted as $H[\cdot]$ operator) approximates the distribution of values, and is a kind of bar graph, where the magnitude of each bar represents the probability that the measured random variable will fall within a given interval (Figure 6 top). Histograms are also often visualized with gray scale maps (Figure 6 bottom). This second method is especially useful if histograms of multidimensional variables are calculated.

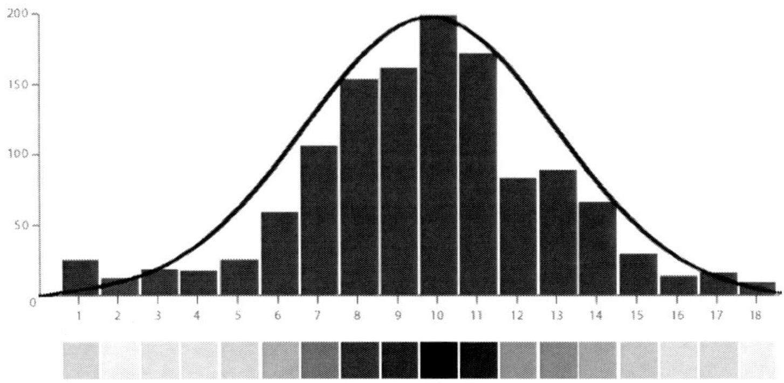

Figure 6. Top: Illustration of exemplary histogram (blue bars), along with the approximated distribution (black curve). Bottom: the same data visualized as gray-scale map.

The histogram is constructed by counting how many values fall into each bin interval. The usual question is: how big should the intervals be? In a digital video system, most often the values of samples $Y_i(\mathrm{x},\mathrm{y})$ are integer values, represented on given bit precision. Classically 8-bit values are used, but more and more often, 10 or even 12 bit samples are used. Thus, a practical choice of bin interval width would be the least significant bit of sample representation, and thus: 1.

On the other hand, what is interesting is that independently from the precision used, the noise values $N_i(x,y)$ that result from such operation (2) are real numbers! That is because the average image $A(x,y)$ is a result of averaging $n+m+1$ values, which does not necessarily have to be an integer. This extra precision that we attain can be used for our advantage, because we can use finer bin interval sizes and thus attain a more detailed histogram. Here we will use bin size equal to $^1/_{16}$ of the least significant bit of sample representation. Therefore, the given value of the noise $N_i(x,y)$ will be counted in the histogram in the bin identified by index $B_i(x,y)$ identified as follows:

$$B_i(x,y) = \lfloor N_i(x,y) \cdot 16 \rfloor, \tag{3}$$

where $\lfloor \cdot \rfloor$ depicts the floor rounding operator.

"Poznan Street" sequence [32]

"Poznan Carpark" sequence [32]

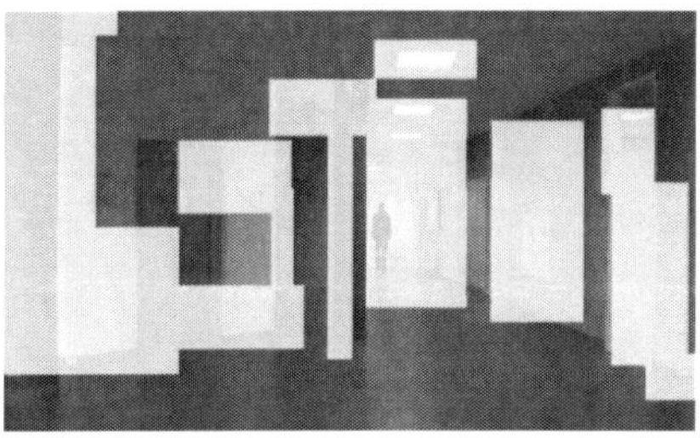

"Poznan Hall" sequence [32]

"Lovebird1" sequence [33]

"Balloons" sequence [35]

"Newspaper" sequence [34]

Figure 7. Regions in the multiview video sequence set that has been manually marked as still for the sake of estimation of the noise. The unused regions have been marked in light-gray.

Now let us see some results! Those will be provided for a set of multiview test sequences [32, 33, 34, 35], attained with the use of mutiview camera systems mentioned in the beginning of the chapter, and with the use of other experimental multiview video systems, during the work of Motion Picture Experts Group (MPEG) of International Organization for Standardization (ISO) and the International Electrotechnical Commission (IEC) on the development of 3D video compression technology.

Table 2. Multiview video sequences used for analysis of the noise

Sequence Name	Frames used	Area of regions marked as still (number of pixels/view/frame)	Camera	Resolution
Poznan Street [32]	32	563 047	Canon XH-G1, 3-CCD camera	1920 × 1088
Poznan Carpark [32]	64	587 030		
Poznan Hall [32]	32	1 384 287		
Lovebird1 [33]	64	105 984	Point Grey Flea camera (CCD), Moritex ML-0813 lenses	1024 × 768
Newspaper [34]	32	71680	Point Grey Research Flea camera with 1/3-inch Sony lenses	
Balloons [35]	48	27 648	XGA CMOS, 8-bit RGB-Bayer camera	

The test material is gathered in Table 2. In all of the used sequences, the range of frames and spatial regions where the scene is still, have been extracted manually. This of course has been done for the sake of simplicity and in real-world applications can be done automatically. The selection of used regions is summarized in Figure 7 (spatial positions of selected regions) and in Table 2 (frame ranges and the area of the marked regions).

3.1. Probability Distributions of Noise Values

The measured probability distributions [30] of the noise are presented in Figures 8-14. Because for most of the cases the measured histograms were identical for all of the cameras, only the averaged results are provided. In general, it can be said that the measured noise distributions are visually very similar to Gaussian (normal) distribution, which has been depicted (dotted-red line) for comparison. The visualized normal distribution has the same parameters: mean value μ and standard deviation σ (Table 3). Interestingly, there are some exceptions for this mentioned "similarity" to the Gaussian distribution, described below.

In the case of *"Poznan Street,"* "*Poznan Carpark*" and *"Poznan Hall"* sequences, the measured distribution is slightly skewed in such a way that the maximum of the distribution is at a position of about 0.4. This may be a result of the internal noise reduction algorithm implemented in the Canon XH-G1 camera or a result of internal non-linear processing of data from the camera sensor. Standard deviations are very similar among the views (Table 3), but there are little differences among the sequences. Those are 2.45 (*"Poznan Street"*), 2.28 (*"Poznan Carpark"*) and 2.01 (*"Poznan Hall"*).

Table 3. Summary of the Gaussian model fitting - parameters of the noise distributions in the test sequences [30]

Sequence Name	Standard deviation	Maximum point of distribution, related to EX	Notes
Poznan Street	2.45	0.41	Measured distribution is skewed
Poznan Carpark	2.28	0.42	
Poznan Hall	2.01	0.51	
Lovebird1, w.o. cam.2	0.66	0.02	Camera 2 of Lovebird1 sequence has vastly different noise profile
Lovebird1, camera 2	1.65	0.01	
Newspaper	1.11	-0.02	-
Balloons	1.01	0.01	-

In the case of the *"Lovebird1"* sequence, the standard deviations are the lowest in the whole test set and are very similar across all of the cameras – at a level of about 0.66. The only exception is camera 2, where the standard deviation is about 2.5 times higher – it has been measured to be about 1.65. This might be evidence that this particular view has been acquainted with different parameters – e.g., the exposure time was shorter, which has been corrected with a higher amplification gain, which also amplified the noise.

Apart from that anomaly, the Gaussians are well-symmetric and centered at the value of 0. This means that the distribution of the noise in such an example is well-centered.

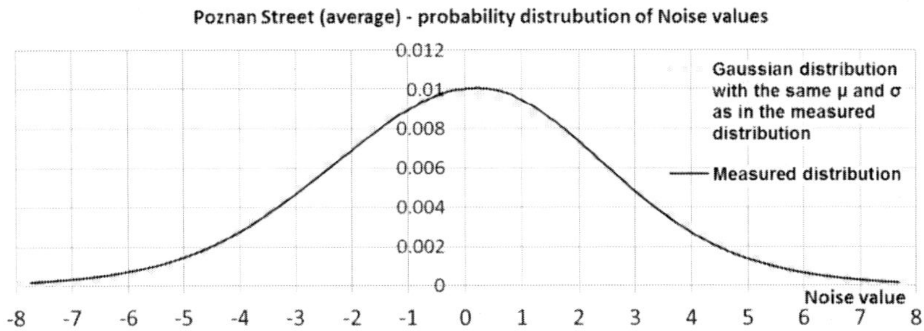

Figure 8. Measured probability distribution of noise values in *"Poznan Street"* sequence (averaged over all views), estimated with histogram bin size of $^1/_{16}$.

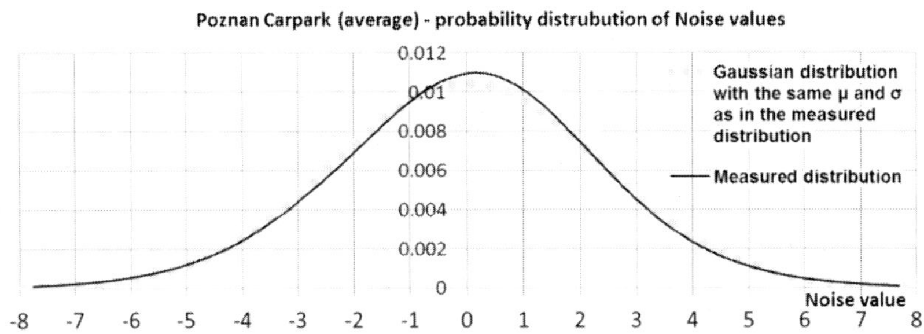

Figure 9. Measured probability distribution of noise values in *"Poznan Carpark"* sequence (averaged over all views), estimated with histogram bin size of $^1/_{16}$.

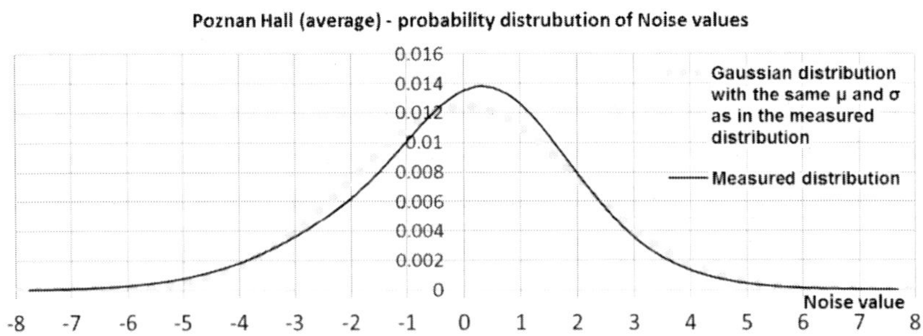

Figure 10. Measured probability distribution of noise values in *"Poznan Hall"* sequence (averaged over all views), estimated with histogram bin size of $^1/_{16}$.

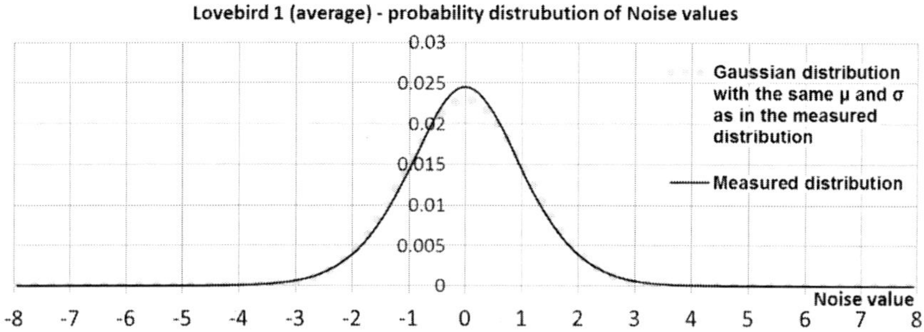

Figure 11. Measured probability distribution of noise values in *"Lovebird 1"* sequence (averaged over all views), estimated with histogram bin size of $^1/_{16}$..

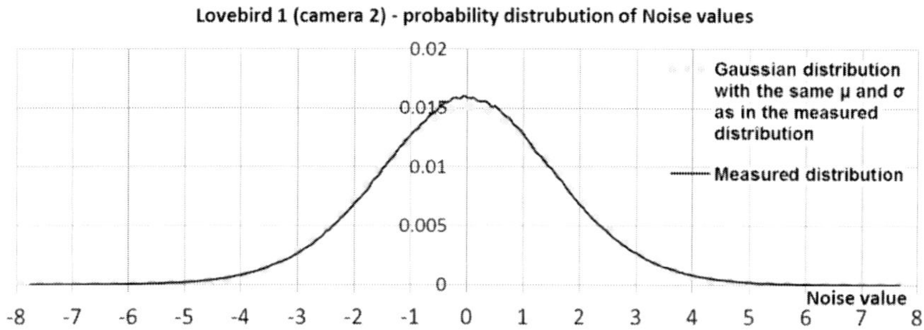

Figure 12. Measured probability distribution of noise values in Lovebird 1 sequence (camera 2), estimated with histogram bin size of $^1/_{16}$. In the case of this camera, the standard deviation is about 2.5 times higher than in other cameras of *"Lovebird 1"* sequence.

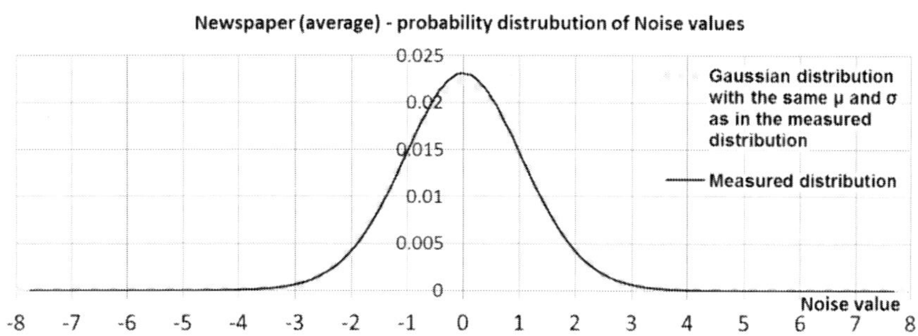

Figure 13. Measured probability distribution of noise values in *"Newspaper"* sequence (averaged over all views), estimated with histogram bin size of $^1/_{16}$.

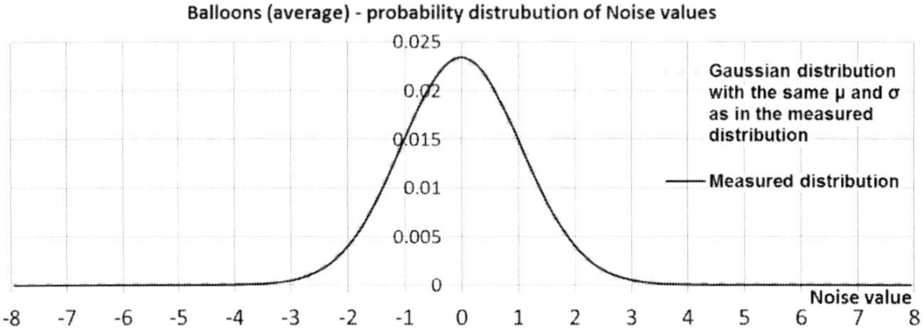

Figure 14. Measured probability distribution of noise values in *"Balloons"* sequence (averaged over all views), estimated with histogram bin size of $^{1}/_{16}$.

The probability distribution of the noise in the *"Newspaper"* sequence is very similar to the Gaussian distribution in all of the cameras. The standard deviations are very similar among the views at a level of about 1.23.

The distributions of the *"Balloons"* sequence strictly follow the Gaussian "bell" shape. Also here, the standard deviations are very similar among the views, at a level of about 1.01.

Although it can be noticed that the measured distributions are visually very similar in shape to the Gaussian function, the visual observation is not statistical proof. Such can be verified with the use of chi^2 goodness-of-fit statistical test. In fact, the following reasoning will show that, in spite of the visual similarity, the measured distributions are not Gaussians: Let's start! Of course we will not go deep into details (surely interesting for the mathematicians), but for the record, we define what are our hyphotheses:

- The working null hypothesis is that the observed distribution is normal (Gaussian).
- The working alternative hypothesis is that the observed distribution is not normal (Gaussian).

For each of the sequences and each of the views we can calculate ${\chi_{gof}}^2$ statistic and compare it relatively to ${\chi_{gof}}^2{}_{critical}$ value:

$$\chi_{gof}{}^{2}{}_{ratio} = \frac{\chi_{gof}{}^{2}}{\chi_{gof}{}^{2}{}_{critical}}. \quad (4)$$

As a right-tailed χ^2 distribution is used, the value of $\chi_{gof}{}^{2}{}_{ratio}$ which is smaller than 1 (measured $\chi_{gof}{}^{2}$ statistic is lesser than $\chi_{gof}{}^{2}{}_{critical}$) means that the null hypothesis cannot be rejected and thus the observed distribution may be Gaussian. Value of $\chi_{gof}{}^{2}{}_{ratio}$ which is greater or equal than 1 (measured $\chi_{gof}{}^{2}$ statistic is greater/equal than $\chi_{gof}{}^{2}{}_{critical}$) means that the null hypothesis must be rejected and thus the observed distribution is not Gaussian.

The results of equation (4) calculated for the test sequences are gathered in Table 4. It can be noticed that for most of the cases, the ratio between $\chi_{gof}{}^{2}$ and $\chi_{gof}{}^{2}{}_{critical}$ is of the magnitude of about $10^1 - 10^2$ proving that the distributions are not Gaussian. The only exception is the Balloons sequence, where $\chi_{gof}{}^{2}{}_{ratio}$ fluctuates around 1 (the presented multiplied showing the level of magnitude of 10^0). Thus, depending on a particular camera of the Balloons, the hypothesis that the distributions are Gaussian must be rejected (marked in white in Table 4) or may not be rejected (marked in gray in Table 4).

Table 4. χ^2ratio results for all views of the tested sequences [30]. Values that are less than 1.0 (marked in gray) indicate that given cases pass the χ^2 test and thus the hyphotesis that the distribution is Gaussian may not be rejected

Sequence Name [Camera indices]	*M*	0	1	2	3	4	5	6	7	8
		χ^2ratio, scaled by the multiplier *M*								
Poznan Street [0..8]	10^1 ×	7.93	7.65	6.71	6.82	7.00	4.90	5.54	5.51	5.11
Poznan Carpark [0..8]	10^2 ×	3.89	3.56	3.03	3.18	3.03	3.33	3.31	2.02	1.89
Poznan Hall [0..8]	10^3 ×	2.12	1.66	1.84	1.75	1.64	2.08	1.76	1.55	1.28
Lovebird1 [0..8]	10^2 ×	0.50	1.49	0.46	1.84	1.95	1.56	1.08	0.86	1.33
Newspaper [0..8]	10^1 ×	1.30	1.38	1.03	2.07	1.92	1.24	2.03	1.84	2.65
Balloons [0..6]	10^0 ×	1.03	1.42	1.16	0.88	0.94	1.90	0.69	-	-

Therefore, in spite of the visual impression that the observed probability distributions are Gaussian-like, generally it can be concluded that for most of the sequences, the null-hypothesis must be rejected and *almost none of them is Gaussian* (at given confidence level). This is an interesting observation because Gaussian distribution of noise values is often assumed seamlessly! Fortunately, as can be seen from the graphs, the error of such an assumption is not significant.

3.2. Independence of the Noise

Another interesting thing that we can check is whether noise values $N_i(x, y)$ are indeed independent. Maybe not, yet they are correlated! For this, we can use a very common tool which is the Pearson Correlation Coefficient. In particular, we can check whether noise values in two successive frames: $i - 1$ and i are correlated. The results of $PCC_{i,i-1}$ are gathered in Table 5. It shows a maximum of absolute value of $PCC_{i,i-1}$, which gives us indiction about the maximal correlation (no matter whether positive or negative), average of absolute value of $PCC_{i,i-1}$ (mean correlation, no matter whether positive or negative), and extreme values of $PCC_{i,i-1}$ (the strongest positive and the strongest negative correlation). As it can be seen, even the highest attained values (like 0.1295 for the *"Balloons"* sequence or about 0.0569 for some of *"Poznan"* sequences) do not provide evidence that the random variables are linearly correlated. Also, the levels of the measured correlation coefficient for other sequences (Table 5) indicate that the noise in subsequent frames is not linearly correlated. On average, the linear correlation coefficient is about ±(0.01-0.03), which is negligible.

Thanks to the experiment above we know that the noise in subsequent frames is not linearly correlated. Of course, linear correlation is the only existing form of dependence between random linear variables. A simple and robust method of verifying whether there is any dependence between two random variables, e.g., α and β is the usage of formula:

$$p(\alpha, \beta) = p(\alpha) \cdot p(\beta), \qquad (5)$$

where $p(\cdot)$ is probability. Such a condition holds true only if such the two variables α and β are independent.

In order to perform such a verification, the two-dimensional histograms of $N_i(x, y)$ versus $N_{i-1}(x, y)$ can been measured. Such a histogram is 2-dimensional and thus it is good to use a visual presentation with gray-levels map as mentioned before in Figure 6. An exemplary histogram has been presented in Figure 15 (top)Fi. One axis of the histograms relates to the values of noise in the frame i, and the other axis relates to the values of noise in frame $i - 1$. The analogous histograms for other sequences are averaged over all frames. Only results for cameras 0-4 have been shown for the sake of brevity, because the remaining ones were looking the same.

If there would be any dependence between those two random variables (modeling noise in frame i and $\mathrm{i} - 1$), there would be an asymmetry in the graph, related to the fact that condition (5) is not meet. Let us check that. As can be noticed (Figure 5), the graphs are separable (5) in both dimensions which indicates that the random variables are independent and therefore the noise in subsequent frames is independent.

Table 5. Linear correlation coefficient calculated over pairs of frames in test sequence set [30]

Sequence Name	$\max_i \lvert PCC_{i,i-1}\rvert$	$\operatorname{average}_i \lvert PCC_{i,i-1}\rvert$	$\min_i PCC_{i,i-1}$	$\max_i PCC_{i,i-1}$
Poznan Street	0.0569	0.0142	-0.0569	0.0344
Poznan Carpark	0.0561	0.0158	-0.0561	0.0533
Poznan Hall	0.0211	0.0084	-0.0211	0.0138
Lovebird1	0.0396	0.0117	-0.0396	0.0339
Newspaper	0.0360	0.0113	-0.0360	0.0293
Balloons	0.1295	0.0325	-0.1163	0.1295

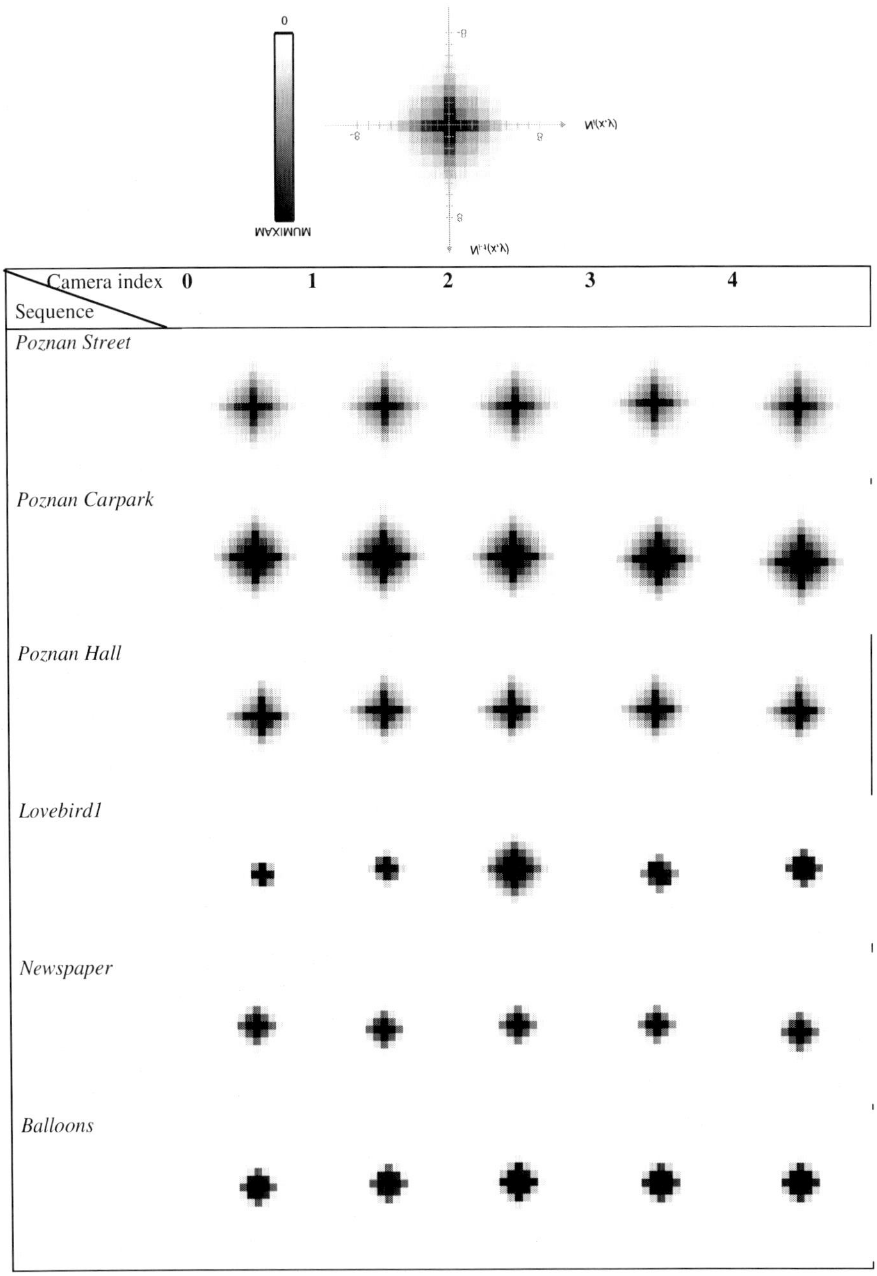

Figure 15. Exemplary zoomed plot (top) and plots of two-dimensional histograms of $N_i(x, y)$ vs. $N_{i-1}(x, y)$ for various test sequences and cameras, averaged over frames. More results can be found in [30].

The same principle that we have used for visual verification can be applied in the mathematical test. The recipe is quite simple. We will need the given normalized 2-dimensional histogram (depicted by $H[\cdot]$ operator) of noise values in the subsequent pair of frames i and $i-1$:

$$H[N_i(x,y), N_{i-1}(x,y)]\,(\alpha,\beta) = h_{i,i-1}(\alpha,\beta), \tag{6}$$

histogram of noise values in frame i:

$$H[N_i(x,y)]\,(\alpha,\beta) = h_i(\alpha) \tag{7}$$

and histogram of noise values in frame i $-$ 1:

$$H[N_{i-1}(x,y)]\,(\beta) = h_{i-1}(\beta). \tag{8}$$

We can assume that those normalized histograms correspond to the probability distributions of noise in the corresponding cases. If the noise distributions are independent between the frames, then according to (5), the expected distribution of $h_{i,i-1}{}'(\alpha,\beta)$ will be:

$$h_{i,i-1}{}'(\alpha,\beta) = h_i(\alpha)\cdot h_{i-1}(\beta)\,. \tag{9}$$

Based on these ingridients, we can perform chi^2 independence test:

$$\chi_{ind.}{}^2 = \sum_{\alpha\in\Phi}\sum_{\beta\in\Phi}\frac{\left|h_{i,i-1}{}'(\alpha,\beta) - h_{i,i-1}(\alpha,\beta)\right|^2}{h_{i,i-1}{}'(\alpha,\beta)}\,. \tag{10}$$

Φ is a set of possible noise values. Range [-8…8] has been selected in order to cover the whole usable range on noise values and at the same time, to avoid small number of samples in histogram bins, which is desired in the case of the chi^2 test. Therefore, $-8 \le \alpha,\beta \le 8$, which results in total $\varphi = 17$. Again, for the record let us note what are the hypotheses for our statistical test:

- The working null hypothesis is that the observed distributions are dependent.
- The working alternative hypothesis is that the observed distributions are independent.

Just like before, for each of the sequences and each of the views, we calculate ${\chi_{ind.}}^2$ statistic and compare it to ${\chi_{ind.}}^2{}_{critical}$ value:

$$ {\chi_{ind.}}^2{}_{ratio} = \frac{{\chi_{ind.}}^2}{{\chi_{ind.}}^2{}_{critical}}. \tag{11} $$

As a left-tailed χ^2 distribution is used, the value of ${\chi_{ind.}}^2{}_{ratio}$ which is greater or equal than 1 (measured ${\chi_{ind.}}^2$ statistic is greater/equal than ${\chi_{ind.}}^2{}_{critical}$) means that the null hypothesis cannot be rejected and thus the observed distributions of the noise values may be dependent. The value of ${\chi_{ind.}}^2{}_{ratio}$ which is smaller than 1 (measured ${\chi_{ind.}}^2$ statistic is lesser than ${\chi_{ind.}}^2{}_{critical}$) means that the null hypothesis must be rejected and thus the observed distributions of the noise values are independent (at the given confidence level).

Table 6. Results of chi^2 independence test, for pairs of succesive frames of the test sequences [30]. The results have been averaged over time and over cameras

Sequence name	${\chi_{ind.}}^2{}_{ratio}$
Poznan Street	0.0145
Poznan Carpark	0.0249
Poznan Hall	0.0194
Lovebird1	0.0387
Newspaper	0.0269
Balloons	0.0307

From the results presented in Table 6, it can be seen that ${\chi_{ind.}}^2{}_{ratio}$ is definitely below 1 (ranges from 0.0145 to 0.0387 which is negligible). This leads to a conclusion that the null hypothesis has to be rejected. Finally, this provides evidence that *the noise in subsequent frames is independent.*

4. Noise-Aware Multiview Processing

It has already been shown that it is possible to gain knowledge about noise from the content itself, and thus without having access to the capture systems and without performing calibration steps e.g., involving color patterns.

Now let us see some examples showing that such a knowledge and thus, being aware of the presence of noise, can be used to improve processing of multiview video. These will include depth map estimation and multiview video compression.

4.1. Depth Map Estimation

Depth map estimation is one of the most crucial problems in multiview video processing. Together with the views, the depth data can be used to create a representation of a 3D scene. Such a representation, called multiview video plus depth (MVD), has recently gained interest among researchers working on delivery formats for autostereoscopic displays, free viewpoint television (FTV), and 3D video compression.

Although many techniques are already known, the depth estimation is still a demanding task with respect to both the quality of the estimated depth and the computational complexity of the algorithms. Moreover, for practical reasons, depth estimation is expected to be fully automatic and work without human supervision. Unfortunately, a variety of state-of-the-art depth estimation algorithms are controlled with parameters that have to be selected manually. Is it possible to *avoid such manual selection of parameters*? Of course there are many research works in this subjects and one of the approaches is to *exploit the knowledge about the noise*. We will consider one such technique below.

In general, depth estimation can be expressed as an optimization problem, which can be solved with the use of generic methods. The most commonly used optimization algorithms are graph cuts (GC) [36] and belief propagation (BP) [37]. The descriptions of those algorithms are off-scope, because we will use them solely as tools for the optimization of depth maps.

The optimization goal is to find the global optimum (not a local optimum), and therefore, the optimization function has to be defined over a whole area of estimated depth map. In one of the commonly used definitions, [36, 37, 38], optimization function f is composed of two components, $DataCost$ and $TransitionCost$ as follows:

$$f \coloneqq \alpha \cdot \sum_p DataCost_p(d_p) + \beta \cdot \sum_p \sum_{q \in N(p)} TransitionCost_{p \to q}(d_p, d_q) \ , \quad (12)$$

where:

f – Optimization function,
p – Each pixel (points) in the estimated depth map,
d_p – The label of the assumed depth of pixel p,
q – Some pixel (point) in neighborhood $N(p)$ of pixel p,
d_q – The label of assumed depth of pixel q,
α, β – Constant value.

$DataCost_p(d_p)$ models direct correspondence between pixels, and express how given pixel p is similar to those pointed to by the depth, represented by d_p in other views of the multiview video.

$TransitionCost_{p \to q}(d_p, d_q)$ penalizes depth maps that are not smooth. If given pixel p has a vastly different depth d_p than its neighbors (pixels depicted as q), it gets a high $TransitionCost$ penalty.

The relation between $DataCost$ and $TransitionCost$ values control the regularization term of the sought depth map; e.g., if we want to get very smooth depth map, $TransitionCost$ values should be considerably higher than $DataCost$ values. Both $DataCost$ and $TransitionCost$ are dimensionless and thus the only way to emphasize one of them is through weighting, which in formula (12) is done with α and β constants. Often only one constant is used (the other one is set to 1) and then due to its role, such a constant is called the *smoothing coefficient* [39]. In many depth estimation techniques, *smoothing coefficient* parameter has to be set manually by the human.

Let us focus on the $DataCost$ component. As it can be found in the literature [30], [39], Maximum A posteriori Probability rule (MAP) can be used to theoretically define $DataCost$ based on the statistical properties of the multiview video and the characteristics of noise, in particular.

In such an approach, we search for the disparity value $d_{x,y}{}^{*}$ that would maximize the probability of $p(d_{x,y})$ under the condition of a posteriori observation of luminance values in both views. This probability will be demarked as $p_{x,y,d}$:

$$d_{x,y}{}^{*} = \underset{arg\ d}{max}\left(p_{x,y,d}\right). \tag{13}$$

Therefore, thanks to MAP rule we can find the optimal depth $d_{x,y}{}^{*}$. We can, if we are able to know what is $p_{x,y,d}$. Fortunately, with some assumptions about the noise, e.g., the presence of additive noise and independence of noise values in the different views of multiview video, and about the scene, e.g., about Lambertian reflectance in the scene, allow to derive a formula for $p_{x,y,d}$.

$$p_{x,y,d} = \frac{p(d_{x,y})}{p(R_{x,y})} \cdot N\left(L_{x+d_{x,y},y} - R_{x,y}\right), \tag{14}$$

where:

$p(d_{x,y})$ is the probability distribution of depth $d_{x,y}$, - this can be approximated with anticipated depth structure of the scene (e.g., more objects in the front or in the background).

$p(R_{x,y})$ is the probability distribution of luminance values, e.g., in the right view. This can easily be calculated by taking a histogram of them in the considered video.

$L_{x+d_{x,y},y} - R_{x,y}$ is the luminance difference between pixels in the left and the right view, matched with an assumed depth value $d_{x,y}$.

$N(\cdot)$ is the probability distribution of the noise in the video.

We already confirmed the independence of the noise and we also know that the probability distribution of noise in the video can be quite easily measured. Interestingly, if we assume that the distribution of the noise is Gaussian (which although turned to be not quite true), and that the probability distributions $p(d_{x,y})$ and $p(R_{x,y})$ are uniform, we can get a simplified (14). Furthermore, if we express the result in a logarithmic scale, we get classical squared differences formula for a pixel similarity metric, allowing us to define $DataCost$:

$$DataCost_p(d_p) = -10 \cdot log(p_{x,y,d}) = -10 \cdot \frac{1}{2\sigma^2}\left(L_{x+d,y} - R_{x,y}\right)^2 \quad (15)$$

where σ is the standard deviation of the assumed Gaussian noise distribution.

The minimization of $-10 \cdot log(p_{x,y,d})$ (with contrast to maximization of $p_{x,y,d}$) can be directly used to find the optimal sought depth ${d_{x,y}}^*$.

Formula (15) can be extended to a block of pixels (instead of single pixel) in order to get a formula reminding *Sum of Squared Differences (SSD)* block matching formula; e.g., for the block of pixels of size $\mu = (2 \cdot S + 1) \times (2 \cdot S + 1)$:

$$SSD_{x,y,d} = \frac{1}{\mu}\sum_{x'=-S}^{S}\sum_{y'=-S}^{S}\left(L_{x+x'+d,y+y'} - R_{x+x',y+y'}\right)^2 \quad (16)$$

SSD is very often used heuristically in depth estimation, but thanks to theoretical derivation based on MAP we know under what mathematical conditions we can use it. Even more importantly, we know what is the scaling factor for the $DataCost$ – it can be derived from the standard deviation of the assumed Gaussian noise!

Now we see how the idea performs in practice. In order to check whether we achieved some gains, we need some kind of reference technique. A very good choice for this is Depth Estimation Reference Software (DERS), developed by ISO/IEC MPEG during works on development of 3D video compression technology. DERS uses the general formulation of optimization function shown in (12). As similarity metric for formulation of $DataCost$, the Sum of Squared Differences (SSD) (16) is used, with *smoothing coefficient* as weighting factor. Therefore, technique used in DERS is supervised in a sense that the *smoothing coefficient* has to be given. Apart from this deficiency, DERS is widely used in scientific literature and here we can do the same.

We use DERS as a reference technique for comparison, and for the sake of fairness, a wide range of smoothing coefficients have been tested. The best and the worst performing settings for each sequence were identified and, for clarity, results only for those values are presented.

Also, we can use DERS as a platform for implementation of the devised technique based on MAP, highlighted above, and described in detail in [30].

Because there are no ground-truth depth maps for the considered multiview sequences (Table 2), it is impossible to directly assess the quality of estimated depth. Therefore, an indirect quality assessment method is used, which is depicted in Figure 16. Three views are considered: the left view, the right view, and the middle view. First, for the left view and for the right view, depth maps are estimated. The estimated depths for both views, along with their original video, are used to synthesize a virtual view in the position of the middle view. The original video of the middle is used as a reference and for comparison, which provides indirect evaluation of the depth map estimation algorithm used. For the sake of the synthesis of virtual views, we use View Synthesis Reference Software (VSRS) [41] also developed by ISO/IEC MPEG.

The overall results are presented in Table 7. The devised depth estimation technique based on MAP is similar to the best case of the original (unmodified) DERS in most cases, and is a little better in some cases.

On average, over the tested sequences, the devised technique provides about a 0.08 dB gain of PSNR (Peak Signal-to-Noise Ratio) versus the best identified case generated with the original (unmodified) DERS (with a manually crafted smoothing coefficient per sequence) and provides about a *2.79 dB gain* of PSNR over the worst case generated by DERS.

Apart from the PSNR gains, the most important thing to note is that the devised depth estimation technique does not require any manual settings (usage of such depth estimations is thus unsupervised). Thanks to the knowledge about the characteristics of noise in multiview video, the parameters employed optimization function model based on MAP rule can be inferred.

Table 7. Objective gains obtained with joint usage of the proposed *DataCost* and *TransitionCost* models [30], [39], related to the best and the worst results obtained with the original (unmodified) DERS, depending on *smoothing coefficient* parameter settings

Sequence Name	Objective gain measure PSNR (Peak Signal-to-Noise Ratio) [dB]: virtual view versus the original view. Virtual view was synthesized based on disparity maps estimated with the use of the following:		
	Original (unmodified) DERS: the worst setting of the *smoothing coefficient*	Original (unmodified) DERS: the best setting of the *smoothing coefficient*	Proposed probabilistic model implemented in DERS
Poznan Street	27.56	31.98	32.02
Poznan Carpark	29.05	30.71	30.95
Poznan Hall	32.17	32.85	32.81
Lovebird1	27.09	29.80	29.83
Newspaper	27.86	31.91	31.95
Balloons	29.95	32.94	32.98
Average	28.94	31.70	31.76
Avg. gain with the proposed method related to the references	+2.81	+0.06	-

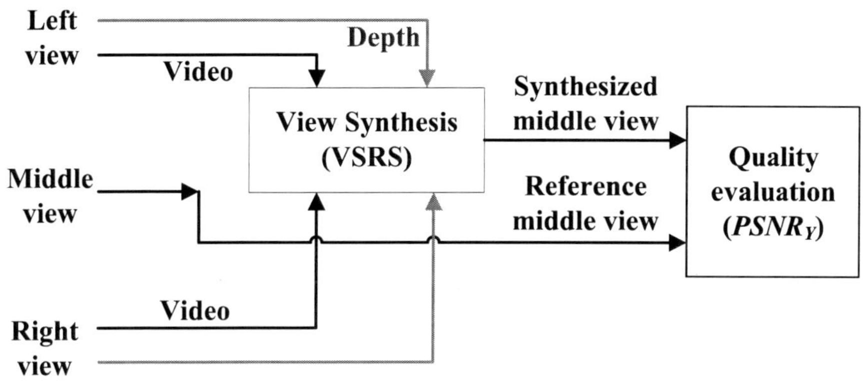

Figure 16. Indirect assessment of depth map quality.

4.2. Multiview Video Compression

One of the most important branches of video processing is video compression. It is even more relevant if multiple views are considered, because the required bitrates are strongly influenced by the number of views. For uncompressed raw bit-streams we can use the following equation:

$$B = V \cdot W \cdot H \cdot C \cdot D \cdot F \tag{17}$$

where B - required bitrate (in bits per second) of a medium, e.g., transmission bandwidth of a connection or bitrate of a file, V – number of views, W and H - width and height (in pixels) of picture, C - number of full-resolution color components (3.0 for 4:4:4 chrominance sub-sampling format, 1.5 for 4:2:0 format), D - bit-depth (8 for typical images) and F – frame rate (in Hz). For example, if we consider uncompressed 8-bit 4:2:0 raw data stream from 9 HD (1920×1080) resolution cameras, with frame rate of 25 frames per second, we get an approximately required bitrate of roughly 6 Gbit/s.

Obviously, efficient compression is needed. The straight-forward solution is to encode each view independently with the use of classical monoscopic video compression technology. Such a solution is often called a simulcast. Although modern compression techniques like HEVC allow for compression ratios above 1000, the resultant bistream still depends linearly on the number of views V.

Another, more interesting solution is to use multiview *compression technology,* like MVC, MV-HEVC (Multiview-HEVC) or 3D-HEVC [42, 43], where many views are coded at the same time with the use of inter-view predictions. Such an approach provides a further significant reduction of bitrate.

How do such matters relate to noise in multiview video systems, which is the subject of this chapter? Independently from the compression technology, the noise which is present in the original video fed to the codec is compressed and transmitted along with the useful content. Because the noise in different views is independent and not correlated, it cannot be predicted and thus cannot be efficiently compressed. Moreover, before the codec tries to predict content with one realization of noise from one view based on the same content from a different view but with different realization of noise, the noise is copied from one view to another. The effect of this is that the noise in decoded, reconstructed video becomes correlated between the views! Figure 17 presents this phenomenon on the example of the "*Poznan Carpark*" sequence. During the encoding, the right view (Figure 17b) has been used to predict (Figure 17c) the content of the left view. During this, along with the content, also the noise in the left view was predicted. The images are presented with strong granular noise for the sake of visualization (Figure 17a without such unnatural noise is provided as a reference), but in real cases the mathematical differences between those noises would be quite small. Often, during compression, strong quantization is used and such small differences are discarded. In such case the content from the right view would be copied to the left view, along with the noise, without adding any residual which would enable one to differentiate between different realizations of noise. As a result, the noise in the right encoded view (Figure 17b), and the left encoded view (Figure 17c) is the same. Original noise in the left view (Figure 17d) has been depicted to show these differences.

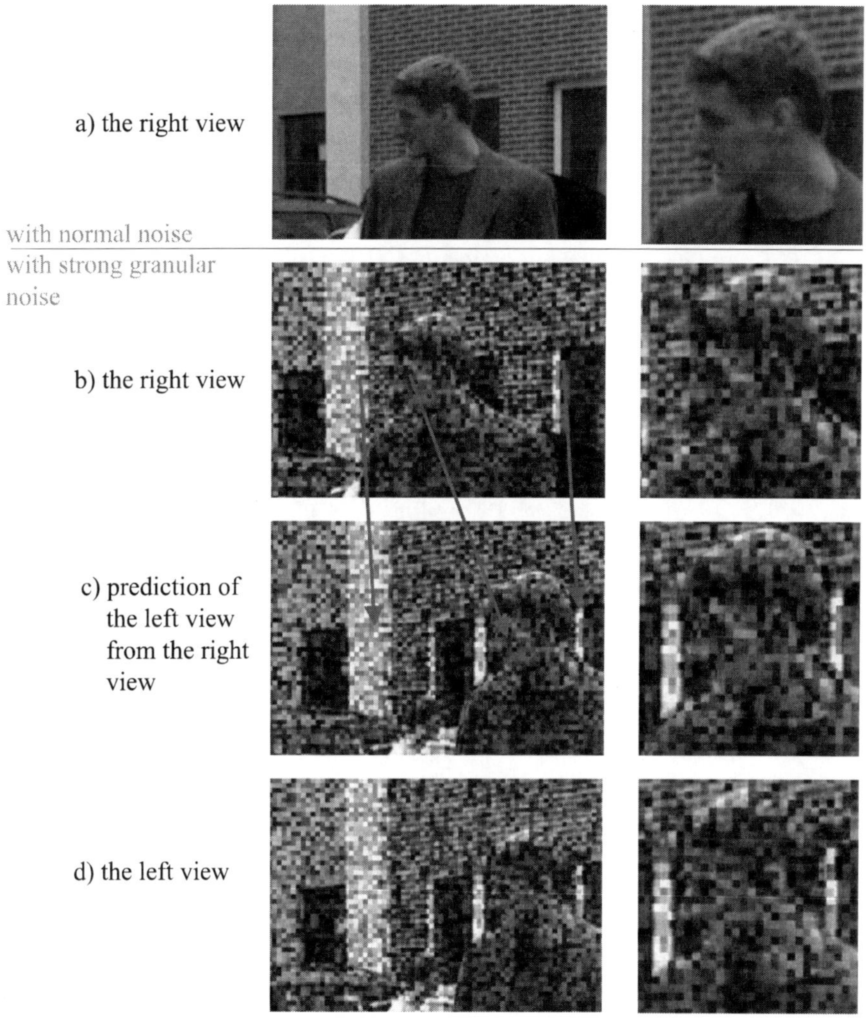

Figure 17. Artifacts related to noise copying by prediction between the views, on the exampe of *"Poznan Carpark"* sequence. Noise in the right view (b) and the left view (c) are identical, while the left view should have individual realization of noise (d). The images (b)-(d) are presented with strong granular noise for the sake of visualization.

The identity of "noise" among different views, especially when observed in 3D, simultaneously with both eyes, on moving sequence, produces a very annoying visual effect. Can we do something about this? Yes, and even more, we can "kill two birds with one stone." Our second "bird" will be *enhancing compression efficiency!*

Let us consider the following scenario: we encode multiview video with the use of a generic multiview codec, e.g., MV-HEVC, 3D-HEVC [42, 43] or similiar (Figure 18). This may include the encoding of depth maps which accompany the image views. The compressed bitstream is delivered to the receiver, decompressed and presented to the viewer with some 3D display. Depending on the particular application and display technology, additional virtual views may be synthesized from the transmitted view + depth pairs. An example may be an autostereoscopic dispay which does not force the viewer to wear any special 3D glasses, but requires dozens of densely positioned views to be provided. Typically, these dozens of views are synthetized basing on a sparse set of transmitted views.

The modification we will add is to perform the transmission in two layers: the content layer and the noise layer (Figure 19). Before the transmission, the input video is preprocessed with a noise reduction technique. As a result, we get a denoised video which is fed to the multiview video encoder, as usual.

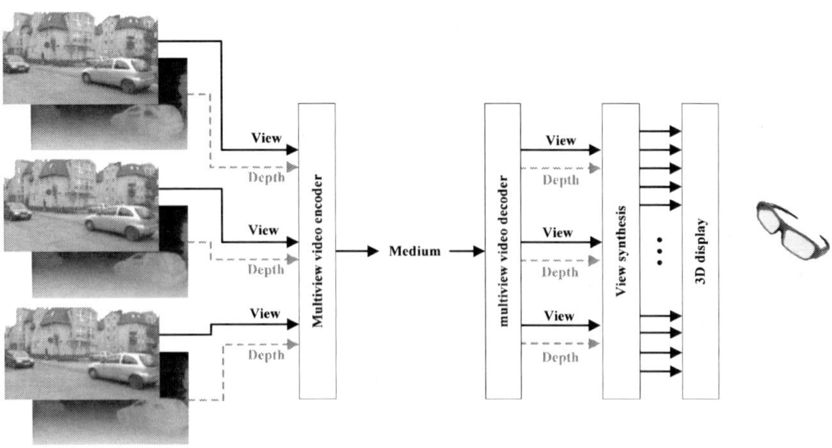

Figure 18. Multiview video delivery scenario.

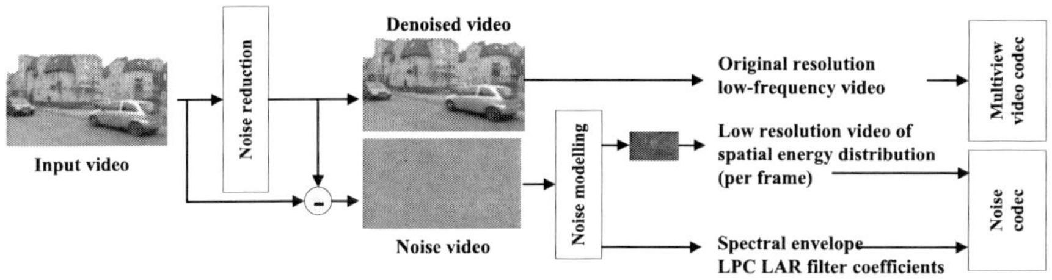

Figure 19. Encoding scheme with noise extraction and modelling.

Of course, the specific separation between the denoised video layer and noise layer depends on the noise reduction technique used. Even though the classification between the content and the noise is fuzzy, we can assume that the residual of noise removal will not usually be compressed efficiently with classic predictive coding. Sample values of this layer are not correlated and random. Therefore, this layer is modeled as a noise.

Can we *more efficiently encode the noise* in that video? Surely – such techniques are known and widely used e.g., in audio compression [44]. Here we can model the noise e.g., as a non-stationary random process. We will use two parameters of this noise model (Figure 19): spatial energy distribution and spectral envelope are coded [45].

Spatial energy distribution is estimated for each frame of noise video. For this purpose, a frame is divided into rectangular non-overlapping blocks. In each of those blocks energy is measured. Energy values, associated with respective blocks, constitute an image of the spatial distribution of energy, whose resolution is smaller than the resolution of the input video. For example, if blocks of size 8×8 are used, the resolution is 8×8 times smaller than resolution of the input video. Spatial energy distribution is coded with use of a HEVC-based coder.

The spectral envelope is estimated from the energy-normalized noise layer with use of a technique similar to Linear Prediction Coding (LPC), known e.g., from audio compression. The result is a set of IIR filter coefficients (in horizontal and vertical direction) which are coded with the use of LAR (log-area-ratio [44]) 8-bit representation. A set of those filter coefficients is estimated for each frame and transmitted to the decoder.

The mentioned noise model can be estimated and delivered to the decoder for each view indepedently, or it can be only one noise model encoded jointly and shared between all of the views. It turns out that the parameters of the noise model are highly correlated among the views, and thus the latter option is more beneficial Frames of the spatial energy distribution of all views are mapped through view synthesis to a position of the base view, and then averaged. This operation results in only one joint spatial energy distribution. Similarly, the energy envelopes of all of the views are averaged, resulting in one joint spectral envelope.

The more interesting stuff happens at the decoder! Because it receives parameters of the model of the noise, it can synthesize the noise in order to give the viewer a natural impression. In particular, a pseudo-random white noise is generated and then modulated by the upsampled spatial energy distribution transmitted in the bitstream and then filtered with IIR filters that resemble the envelope of the original high-frequency layer spectrum. The resultant video, which resembles the noise video extracted originally at the encoder, is added to the reconstructed video in order to restore the noise component.

How does such a method perform? And how can we even tell that quantitatively, when a substantial part of the reconstructed video – noise - is in fact synthesized and thus we cannot use classical objective quality measures like PSNR (Peak Signal-to-Noise Ratio)? The answer is – through *subjective evaluation*. It has been performed as follows.

The same sequence set for the noise analysis above has been used (Table 2). From each sequence, 3 views (texture with depth maps) have been encoded and reconstructed at the decoder. Based on the decoded data, two virtual views have been synthesized exactly in between the original views. For the view synthesis algorithm, we again used MPEG VSRS (View Synthesis Reference Software) [41]. Each synthesized pair of virtual views was presented to the viewers who were asked to assess their quality and vote in 11-point discrete Mean Opinion Score (MOS) scale [46], from 0 meaning "bad" to 10 meaning "excellent." All of this has been performed for three alternative coding methods:

- HEVC simulcast,
- MV-HEVC, and
- MV-HEVC with the devised noise layer modelling.

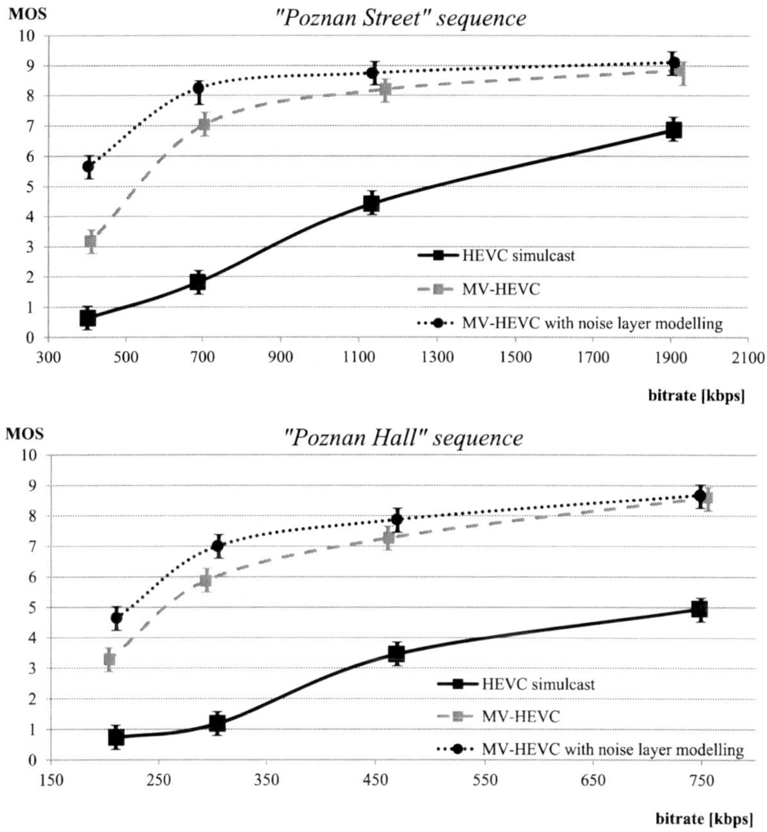

Figure 20. The results of subjective evaluation, performed in Mean Opinion Score (MOS) scale, along with marked 95% confidence intervals. 0 is “bad,” 10 is “excellent.”

For each sequence, four bitrates were used, equal for all of the codecs in order to allow fair competition, which related to four different quality levels. The subjective tests have been carried out in accordance with the Double Stimulus Method based on the general rules of ITU Recommendation BT.500 [46]. A total of about 60 young persons were viewing each stereopair on a 46” Hyundai S465D polarization monitor.

The results are presented in Figure 20 for two selected representative sequences. As it can be seen, for low bitrates the usage of devised noise layer modelling brings quality *gains up to even 2.5 MOS points.* For high bitrates, the gain decrease, and for the highest bitrates the differences become statistically insignificant as the confidence intervals overlap.

On average, we can say that the usage of noise layer modelling is *about 1 MOS point.* Figure 20 also depicts the results of the HEVC Simulcast, provided as reference. Thanks to this, we can see that the usage of a multiview extension of HEVC (MV-HEVC) brings a quality gain of about 2-3 MOS points over the HEVC simulcast. Therefore we can say that such a simple trick as noise extraction and modelling provides on average about one-third of a quality gain provided by dedicated compression technology! If we consider maximal per case gains, they are even more comparable. And all of this is thanks to the awareness of the noise!

CONCLUSION

In this chapter the importance of noise in the context of multiview camera systems has been highlighted. It has been shown how to simply analyze and measure characteristics of noise present in multiview video content. Then, it has been shown how the awareness of noise present in the multiview video can be used to enhance the processing algorithm.

On the example of the depth estimation technique, it has been shown that knowledge about nosie can be used to define the optimization function and to avoid manual selection of control parameters. It has been shown, that with the devised approach, a substantial gain of quality of the estimated depth can be attained, e.g., up to 2.8 dB of objective quality measure (luminance PSNR -Peak Signal-to-Noise Ratio). It is almost twice as small an error and thus a two times higher quality!

In the field of multiview compression, it has been shown that noise present in a multiview video can be extracted, modelled and coded jointly among the views. The advantages of such an approach includes a lack of artifacts coming from noise copying by prediction between the views and higher compression efficiency. Such an approach provides higher quality, about 1, up to 2.5 Mean Opinion Score (MOS) points in subjective tests, while maintaining the same bitrate as Multiview-HEVC (MV-HEVC). For comparison, the gain of MV-HEVC versus HEVC Simulcast has been measured as about 2-3 MOS points. Thus, a very simple tool exploiting knowledge about noise can give a very comparable quality gain on dedicated compression technology, which, with significant complexity shows an increase of the coding. Isn't it amazing?

The conclusion of this chapter is that knowledge about noise present in multiview video can lead to significant enhancement. Hopefully, this will motivate the reader to individually pursue this idea and find more interesting examples on their own.

REFERENCES

[1] C. Fehn, R. de la Barr´e, S. Pastoor, "Interactive 3-DTV – concepts and key technologies," *Proc. IEEE,* Vol. 94, No. 3, pages 524–538, Mar. 2006.

[2] D. Scharstein, R. Szeliski, "A taxonomy and evaluation of dense two-frame stereo correspondence algorithms," *Int. Journal of Computer Vision* 2002.

[3] D. Tian, P.-L. Lai, P. Lopez, C. Gomila, "View Synthesis Techniques for 3D Video," *Digital Image Proc. XXXII. Proceedings of the SPIE,* Vol. 7443, 2009.

[4] S. Farsiu, D. Robinson, M. Elad, P. Milanfar, "Fast and Robust Multi-frame Super-resolution," *IEEE Trans. on Image Processing,* vol. 13, no. 10, pp. 1327–1344, Oct. 2004.

[5] T. E. Bishop, P. Favaro. "The light field camera: Extended depth of field, aliasing, and super-resolution" *IEEE Trans. on Pattern Analysis and Machine Intelligence*, v. 34, 2012.

[6] T. Georgiev, G. Chunev, A. Lumsdaine, "Super-resolution with the focused plenoptic camera," *SPIE Electronic Imaging*, Jan 2011.

[7] M. Joachimiak, D. Rusanovskyy, M. M. Hannuksela, M. Gabbouj "Multiview 3D Video Denoising in Sliding 3D-DCT Domain," *European Signal Processing Conference, EUSIPCO 2012*, Budapest, Romania, 2012.

[8] J. Sun, N. N. Zheng, H. Y. Shum, "Stereo Matching Using Belief Propagation," *IEEE Trans. on Pattern Analysis and Machine Intelligence,* Vol. 25, Issue: 7, pages 787-800, 2003.

[9] S. V. Vaseghi, *"Advanced Digital Signal Processing and Noise Reduction* (Third Edition)," John Wiley & Sons 2006, ISBN: 978-0-470-09495-2, 2006.

[10] P. Leclercq, J. Morris, "Robustness to noise of stereomatching," in *Proc. of the 12th Int. Conf. on Image Analysis and Processing*, pp. 606–611, Sept. 2003.

[11] G. Varghese, Z. Wang, "Video denoising based on a spatio-temporal gaussian scale mixture model," *IEEE Transactions on Circuits and Systems for Video Technology,* vol. 20, no. 7, pp. 1032–1040, July 2010.

[12] C. Boncelet "Image Noise Models" In Alan C. Bovik. *"Handbook of Image and Video Processing."* Academic Press. ISBN 0-12-119792-1, 2005.

[13] "ST 259:2008: *For Television — SDTV1 Digital Signal/Data — Serial Digital Interface*," doi:10.5594/S9781614824077, ISBN 978-1-614 82-407-7, 2008.

[14] https://en.wikipedia.org/wiki/LANC, online 01.2018.

[15] M. Tanimoto, "Ray capture systems for FTV," *Signal & Information Processing Association Annual Summit and Conference (APSIPA ASC),* 2012 Asia-Pacific, 978-1-4673-4863-8, 2012.

[16] F. Zilly *"Method for the Automated Analysis, Control and Correction of Stereoscopic Distortions and Parameters for 3D-TV Applications,"* Doctoral Thesis, August 2015.

[17] I. Feldmann, M. Mueller, F. Zilly, R. Tanger, K. Mueller, A. Smolic, P. Kauff, T. Wiegand, *"HHI Test Material for 3D Video,"* m15413, France, Archamps, 21 April, 2008.

[18] M. Domański, K. Klimaszewski, J. Konieczny, M. Kurc, A. Łuczak, O. Stankiewicz, K. Wegner, "An experimental Free-view Television System," *1st International Conference on Image Processing & Communications (IPC),* Bydgoszcz, Poland, pp. 175-184, September 2009.

[19] M. Domański, A. Dziembowski, D. Mieloch, A. Łuczak, O. Stankiewicz, K. Wegner, "A Practical Approach to Acquisition and Processing of Free Viewpoint Video," *31st Picture Coding Symposium PCS 2015*, Cairns, Australia, pp. 10-14, 31 May - 3 June 2015.

[20] O. Stankiewicz, M. Domański, A. Dziembowski, A. Grzelka, D. Mieloch, J. Samelak, "A Free-Viewpoint Television System for Horizontal Virtual Navigation," *IEEE Transactions on Multimedia,* Volume: PP, Issue: 99, 10.1109/TMM.2018.2790162, January 2018.

[21] H. Tian. *"Noise Analysis in CMOS Image Sensors."* A Dissertation of Stanford University, 2000.

[22] R. C. Gonzalez, R. E. *Woods "Digital Image Processing"* 2nd Ed. Bovik: Elsevier Academic Press, 2002.

[23] K. Arakawa "Median filter based on fuzzy rules and its application to image restoration," *Fuzzy Sets Syst.*, 1996.

[24] M. Lang, H. Guo, J. E. Odegard; C. S. Burrus, R. O. Wells, "Noise reduction using an undecimated discrete wavelet transform", *Signal Processing Letters, IEEE* Volume 3, Issue 1, Jan. 1996.

[25] P. R. Schrater, D. C. Knill, E. P. Simoncelli, "Mechanisms of visual motion Detection," *Nature America*, 2000.

[26] T. Wittkop, V. Hohmann, B. Kollmeier, "Noise reduction strategies employing interaural parameters," *Journal of The Acoustical Society of America* - Vol. 105, No. 2, 1999.

[27] D. Kersten, "Statistical efficiency for the detection of visual noise", *Vision Research*, Volume 27, Issue 6, Pages 1029–1040, 1987.

[28] W. C. Van Etten, *"Introduction to Random Signals and Noise,"* John Wiley & Sons 2006, ISBN 0-470-02411-9, 2006.

[29] R. Dugad, N. Ahuja "Video denoising by combining Kalman and Wiener estimates," *Proceedings of International Conference on Image Processing*, pages 152-156, 1999.

[30] O. Stankiewicz, *Stereoscopic depth map estimation and coding techniques for multiview video systems*, PhD Dissertation at Poznan University of Technology, Faculty of Electronics and Telecommunications, 2014.

[31] O. Stankiewicz, M. Domański, K. Wegner, "Analysis of noise in multi-camera systems," *3DTV Conference 2014*, Budapest, Hungary, 2-4 July 2014.

[32] M. Domański, T. Grajek, K. Klimaszewski, M. Kurc, O. Stankiewicz, J. Stankowski, K. Wegner, *"Poznań Multiview Video Test Sequences and Camera Parameters,"* ISO/IEC JTC1/SC29/WG11 Doc. M17 050, Xian, China, October 2009.

[33] Gi-M. Um, G. Bang, N. Hur, J. Kim and Yo-S. Ho, *"Video Test Material of Outdoor Scene,"* ISO/IEC JTC1/SC29/WG11, MPEG/ M15371, Archamps, France, April 2008.

[34] Yo-Sung Ho, E. K. Lee, C. Lee "Video Test Sequence and Camera Parameters," *ISO/IEC MPEG M15419*, Archamps, France, April 2008.

[35] M. Tanimoto, T. Fujii, N. Fukushima, *"1D Parallel Test Sequences for MPEG-FTV,"* MPEG M15378, Archamps, France, April 2008.

[36] Y. Boykov, O. Veksler, R. Zabih, "Fast approximate energy minimization via graph cuts," *IEEE Transactions on Pattern Analysis and Machine Intelligence,* Vol. 29, pages 1222 1239, 2001.

[37] M. F. Tappen, W. T. Freeman, "Comparison of graph cuts with belief propagation for stereo, using identical MRF parameters," *IEEE International Conference on Computer Vision*, 2003.

[38] Li Hong, G. Chen, "Segment-based stereo matching using graph cuts," *Proceedings of the 2004 IEEE Computer Society Conference on Computer Vision and Pattern Recognition, 2004.* CVPR 2004, Vol.:1, pages 74-82, 27 June-2 July 2004.

[39] M. Tanimoto, T. Fujii, K. Suzuki, "*Video Depth Estimation Reference Software (DERS) with image segmentation and block matching,"* ISO/IEC JTC1/SC29/WG11 Doc., M16092, Lausanne, Switzerland, Feb. 2009.

[40] O. Stankiewicz, K. Wegner, M. Domański, "Depth estimation based on maximization of A posteriori probability," *International*

Conference on Computer Vision and Graphics ICCVG 2016, Warsaw, Poland, 19-21 September 2016,

[41] *"View synthesis algorithm in view synthesis reference software 3.0 (VSRS3.0),"* ISO/IEC JTC1/SC29/WG11 Doc. M16090, Feb.2009.

[42] *"High Efficiency Video Coding,"* ISO/IEC IS 23008-2, 3rd Ed., October 2017, and ITU-T Rec. H.265, 4th Ed., December 2016.

[43] G. Tech, Y. Chen, K. Müller J.-R. Ohm, A. Vetro, Y.-K. Wang, "Overview of the multiview and 3D extensions of high efficiency video coding," *IEEE Transactions on Circuits and Systems for Video Technology,* vol. 26, No. 1, January 2016, pp. 35-49, 2016.

[44] J.A. Martins "Low bit rate LPC vocoders using vector quantization and interpolation," *International Conference on Acoustics, Speech, and Signal Processing,* 1991, vol. 1 pp. 597- 600. 1991.

[45] M. Domański, O. Stankiewicz, K. Wegner, M. Kurc, J. Konieczny, J. Siast, J. Stankowski, R. Ratajczak, T. Grajek, High efficiency 3D video coding using new tools based on view synthesis," *IEEE Transactions on Image Processing*, Vol. 22, No. 9, pp. 3517-3527, Sept. 2013.

[46] *"ITU-R BT.500-12 Recommendation: Methodology for the subjective assessment of the quality of television pictures,"* approved Jan. 2012.

BIOGRAPHICAL SKETCH

Olgierd Stankiewicz

Affiliation: Poznan University of Technology, Poland

Education: Received his PhD from the Faculty of Electronics and Telecommunications, Poznań University of Technology in 2014.

Business Address: Poznan University of Technology, Poznań, Poland

Research and Professional Experience:

Assistant professor at the Chair of Multimedia Telecommunications and Microelectronics, where he has been working since receiving his Master of Engineering degree in 2006. Actively involved in ISO standardization activities where he contributes to the development of the 3D video coding standards. In years 2011-2014, a coordinator of development of MPEG reference software for 3D-video coding standards based on AVC. Now contributes to MPEG Free viewpoint TV and JPEG-PLENO standardization activities. Published over one hundred MPEG/JPEG standardization documents as well as about thirty papers on free view television, depth estimation, view synthesis and hardware implementation in FPGA. His professional interests include signal processing, video compression algorithms, computer graphics and hardware solutions.

In: Noise Reduction
Editor: Marc Farrow

ISBN: 978-1-53613-541-1

Chapter 3

Noise Reduction Techniques in Fiber Optic Sensors

Frederick Seng*[*], *Nikola Stan, Rex King and Stephen Schultz
Department of Electrical and Computer Engineering,
Brigham Young University, Provo, Utah, US

Abstract

Test and evaluation needs are becoming more and more demanding on the sensors being used. Fiber optic sensors are known for their ability to survive in harsh environments such as high pressure and temperature like those within oil/gas wells, or high vibration like those with electromagnetic railguns. As these sensors become commercially available, one challenge is overcoming multiple noise sources simultaneously in the sensor and its network imposed from the harsh environment.

Optical amplifiers are typically used in fiber optic sensors to improve the signal-to-noise ratio (SNR). This method is suitable for boosting the signal above the floor caused by electrical noise. However, certain noise sources in fiber optic sensors directly modulate the optical signal. This can

[*] Corresponding Author Email: sengfrederick@gmail.com.

be due to other measurands being induced on the system. For example, while a slab coupled optical sensor (SCOS) is sensitive to electric field, it is also inherently sensitive to strain; therefore, any strain noise imposed on the sensor would overwhelm the electric field measurement and be considered noise.

In this chapter, optical phase induced interference noise (PIIN) in a sensor network is reduced by using optical phase modulation in both a SCOS and a fiber Bragg grating (FBG). Strain induced noise on the optical fiber sensor is reduced using a differential sensor setup called a push-pull SCOS. Phase modulation and differentiation are then combined to allow for simultaneous noise reduction from different noise sources and allow for optical sensing of electric fields in harsh environments. These principles are applicable to other types of fiber optic sensors and sensors in general.

Keywords: fiber optic sensors, noise reduction, phase modulation

1. INTRODUCTION

With the rise of modern infrastructure and systems, testing and evaluation (T&E) of specific components such as structural health monitoring is becoming more important. Many different sensor types such as acoustical sensors [1], temperature sensors [2], and many more [3, 4 5] have been developed to test and evaluate different types of systems. One common difficulty with the implementation of these modern sensors is noise. Sometimes noise like optical carrier phase fluctuations which are internal to the measurement system can be reduced or eliminated via a more robust sensor setup, or can simply by neglected by improving the SNR of the sensor system [6, 7, 8]. But sometimes the noise caused by events such as vibration external to the system can manifest itself as a measurand on the sensor that overhwhelms the actual measurement [9, 10]. In such cases more clever methods are needed to reduce or eliminate noise in the sensor system. This chapter illustrates how to reduce different noise sources simultaneously to allow for a sensor to operate in a harsh environment. Although the application is illustrated on a fiber optic sensor, the concepts are applicable to a variety of sensor systems.

Fiber optic sensors are ideal for T&E applications due to numerous advantages such as their compact, dielectric and lightweight nature. Figure 1 shows a commercially available electric field sensor called the D-dot, which is very sensitive, however it's large physical size and metallic nature disrupts measured fields. Testing systems with tighter spaces does not allow for the use of this sensor. Figure 2 shows a fiber optic variant of the D-dot called a slab coupled optical sensor (SCOS). The SCOS has a sensing region of 1 mm parallel to the direction of the fiber and 0.3 mm perpendicular to the length of the fiber, allowing it to fit into very tight spaces. It's dielectric nature is also renders it non intrusive to the electric field being measured.

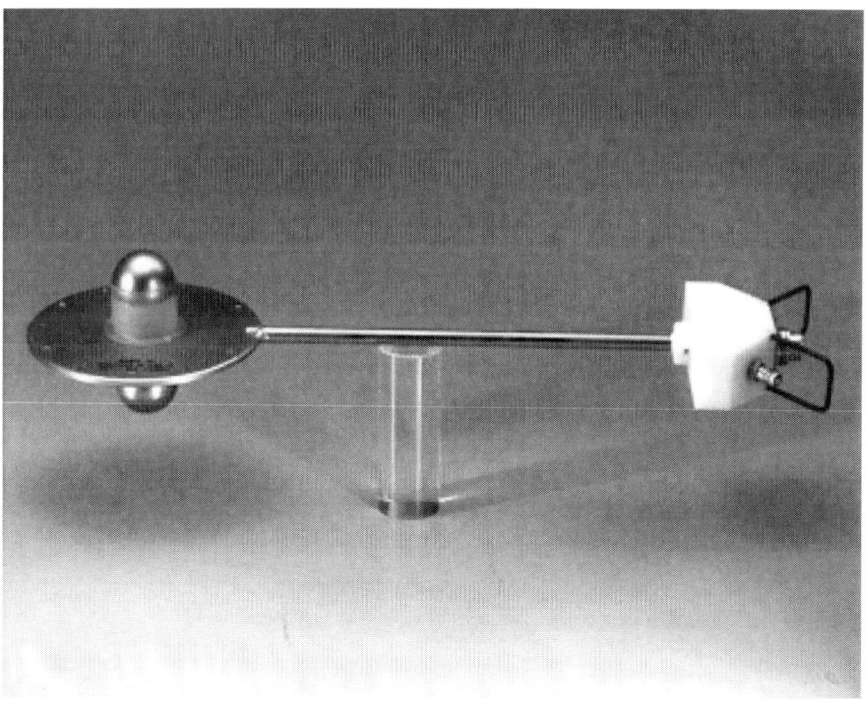

Figure 1. Commercially available D-dot sensor [11].

Despite numerous advantages over their semiconductor/metal counterparts, fiber optic sensors still have a considerable amount of research that needs to be done to improve performance and applicability. The sensor this work uses to illustrate noise reduction techniques is the electric field fiber optic sensor called a slab coupled optical sensor (SCOS) under operation in a harsh environment.

Figure 2. The Slab Coupled Optical Sensor (SCOS) has a measurement cross section of 1 mm along the length of the fiber and 0.3 mm perpendicular to the length of the fiber.

A new differential push-pull variant of the SCOS is made to subtract out noise, and phase modulation is used to reduce interferometric noise. By combining these two noise reduction methods it is possible for the SCOS to measure electric fields in a harsh environment. In otherwords, this chapter shows a method to simultaneously take out different noise types in a SCOS to handle a harsh environment. These concepts are applicable to other types of sensors as well, where different noise sources need to be reduced simultaneously to recover the measured signal.

2. The Slab Coupled Optical Sensor (SCOS)

To understand the noise reduction techniques highlighted in this chapter it is important to first understand the operation of the sensor these techniques are illustrated on. The performance of many systems can be characterized by measuring the electric field they produce under operation. The compact dielectric nature of fiber optic electric field sensors are ideal for measuring the electric field signals of these systems [11, 12, 13, 14]. Unfortunately, many of these systems also produce multiple external noise types or even amplify internal noise types which contribute to a harsh environment and can overwhelm the measurement obtained by the fiber optic sensor [15]. In order to overcome these harsh environments the fiber optic sensor must be able to simultaneously overcome all the different noise types from a system.

Examples of systems that need to have their electric field characterized and produce harsh environments on fiber optic sensors are rail guns which produce enormous amounts of vibrational and acoustic noise under operation [16, 17], and plasma discharge systems for combustion ignition which produce a significant amount of vibrational and RF noise [18].

To understand how the noise sources affect the SCOS its operation must first be understood. A variant of the slab coupled optical sensor (SCOS) [19, 20, 21, 22] called the push-pull SCOS [22] is used in this work. Figure 3 shows that the SCOS consists of a lithium niobate crystal adhered to a polarization maintaining D-shaped optical fiber. It is crucial that the D-fiber is polarization maintaining since the sensitivity of the SCOS depends on interrogating the index change in the lithium niobate crystal through the r_{33} electro optic coefficient.

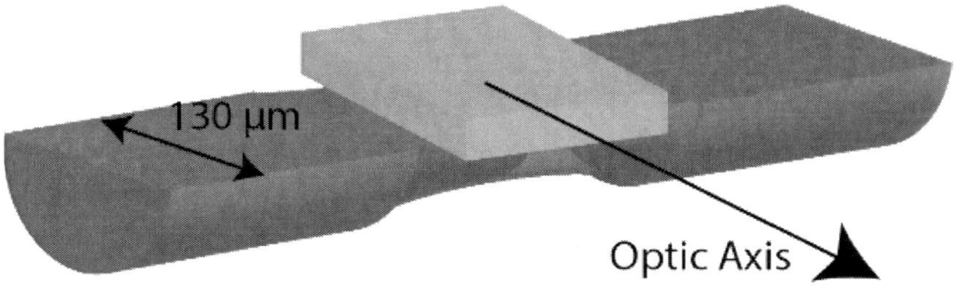

Figure 3. The Slab Coupled Optical Sensor (SCOS) consists of a lithium niobate crystal adhered to a D-shaped optical fiber.

The D-shaped optical fiber has a section where its flat region is etched down close to the elliptical polarization maintaining core via hydrofluoric acid for better coupling between the crystal and the fiber. When the crystal is adhered onto the fiber, specific wavelengths of transverse polarized light in the direction of the lithium niobate optic axis can couple out of the fiber into the crystal as given by [11]

$$\lambda_m = \frac{2t}{m}\sqrt{n_o^2 - N_f^2}\,, \tag{1}$$

where t is the thickness of the slab waveguide, n_o is the refractive index of the slab waveguide, N_f is the effective index of the fiber mode, and m is the slab waveguide mode number.

Figure 4 shows that when certain wavelengths of light are coupled out of the D-fiber, that resonance dips form in the transmission spectrum of the optical fiber located at λ_m. An electric field applied in the direction of the optic axis of the lithium niobate crystal will change the refractive index of the slab waveguide with respect to the r_{33} coefficient.

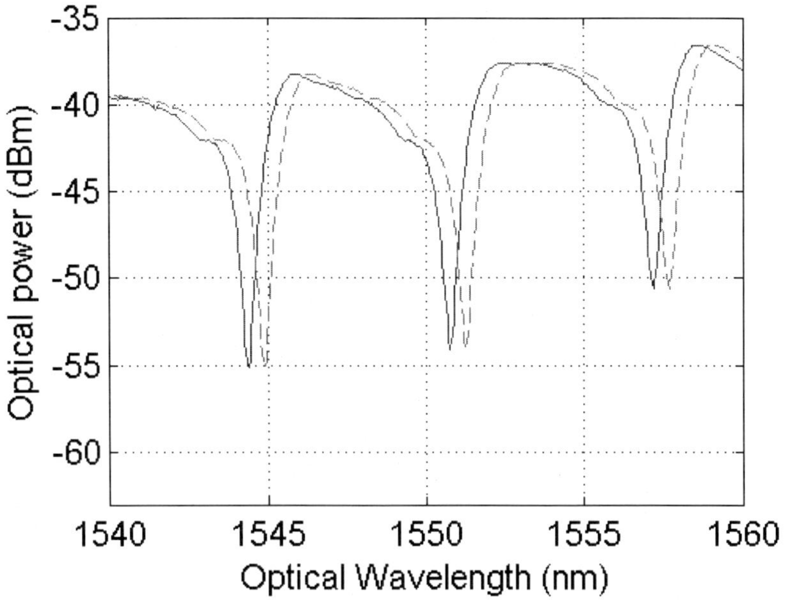

Figure 4. Transmission spectrum of the SCOS (solid) without and (dashed) with an applied electric field.

Equation (1) states that due to a change in the refractive index of the lithium niobate crystal the resonance dips shift. This shift is illustrated in Figure 4 and modulates the power of a laser launched into the D-fiber that is aimed at a resonance edge. Therefore monitoring the change in the power transmitted through the fiber allows for the electric field applied across the optic axis of the lithium niobate crystal to be deduced. Unfortunately, Equation (1) also states that random changes to N_f can also shift the resonance edge. Changes to N_f can be caused due to strain on the fiber from localized vibration noise. This localized vibration noise and RF noise can be reduced by using a differential version of the SCOS called the push-pull SCOS [22].

3. Differential Method for Localized Vibrational Noise Reduction

Figure 5 shows an example of the measured electric field from a SCOS. The (top) shows the actual applied electric field to the SCOS and the (bottom) shows the measured voltage on the Oscilloscope. The measured voltage is related to the actual applied electric field by a calibration constant. This calibration constant is obtained by measuring a known applied electric field and relating that known field to the measured Oscilloscope voltage. As long as the measured electric fields are not too high, the amount of applied electric field can be assume to have a linear relationship with the measured voltage on the Oscilloscope [12]. Periodic exponentials are used in this work because periodic transient signals span a wide range of frequencies which does not allow for traditional filtering methods to reduce noise within a system.

Figure 6 shows a measured signal from a SCOS in a high vibration environment. The applied electric field across the SCOS is the same as the applied electric field in Figure 5, but it is unrecognizable due to the localized vibration noise induced on the SCOS. A known method for reducing external noise in sensors is by using a differential setup [9], where 2 sensors are configured such that the external noise is measured with the same magnitude sign, but the desired measurand is measured with opposite magnitude signs. Two measurements are acquired at the same time from both sensors allowing for a subtraction between the two measurements to differentiate out the noise and recover the desired measurand.

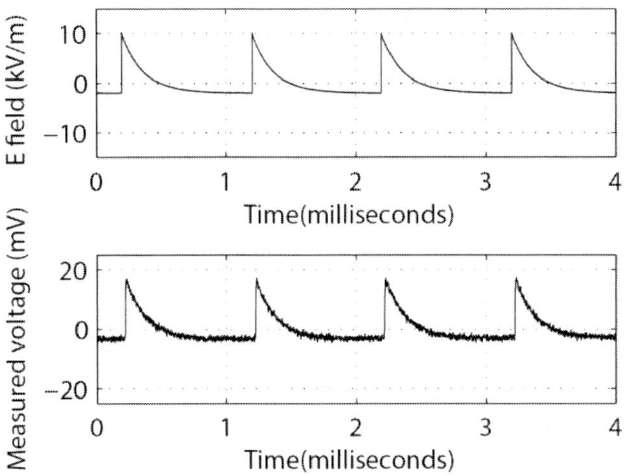

Figure 5. (Top) Applied electric field and (bottom) measured voltage.

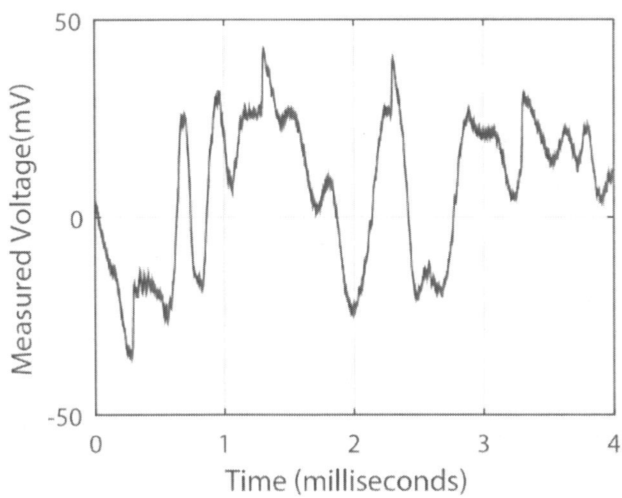

Figure 6. SCOS signal measured with an impact next to the SCOS.

As a result a new variant of the SCOS called a push-pull SCOS is developed and used in order to discern the electric field signal from the noise. The push-pull SCOS consists of 2 SCOS on the same fiber with their optic axes flipped opposite to each other to produce 2 simultaneous measurements, one from each SCOS. Because of this configuration the electric field is measured from both sensors in opposite magnitude signs, but the strain induced noise from localized vibrations is measured with the same magnitude sign. This allows for a subtraction of the two signals to recover the measured electric field signal from the SCOS.

3.1 Push-Pull Sensor Overview

Figure 7 shows an illustration of the push-pull SCOS. The two crystals are adhered on the fiber with their optic axes flipped 180 degrees from each other. This way the 2 simultaneous measurements from the 2 SCOS measure strain with the same magnitude sign but measure the electric field with opposite magnitude signs, allowing for a subtraction between the two signals to differentiate the noise. In theory, this configuration should also allow for reduction in thermal noise as well.

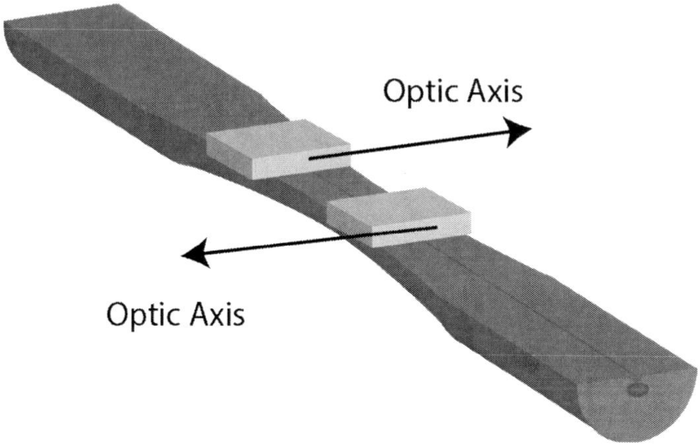

Figure 7. A push-pull SCOS consists of 2 lithium niobate crystals on a single D fiber with their optic axis flipped opposite with respect to each other.

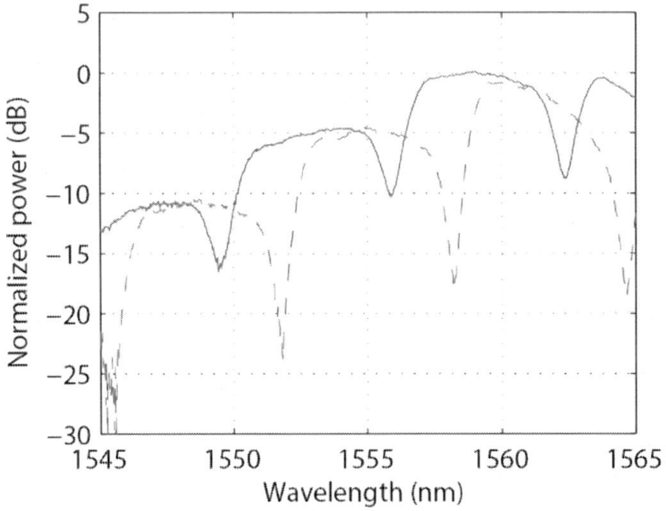

Figure 8. Normalized logarithmic SCOS transmission spectrum with (solid) Lithium Niobate crystal and (dashed) Lithium Niobate Crystal with a layer of Silicon Nitride.

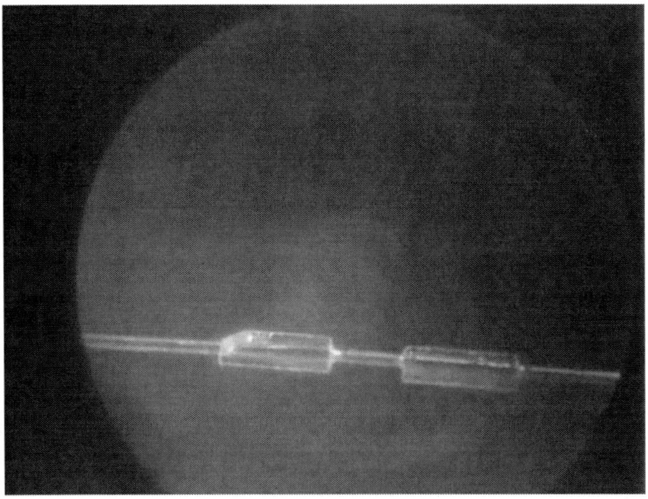

Figure 9. A Push-Pull SCOS with 2 Lithium Niobate Crystals on a Single D-fiber. One crystal is notched in the upper left corner to identify which crystal has been altered with silicon nitride.

To multiplex the two different sensors on the push-pull SCOS, it is important to make sure that the two interrogated signals don't interfere with each other. This is the same for multiplexing any differential sensor setup, it is crucial that the signals from the two sensors do not interfere with one another. In the case of the push-pull SCOS this means using 2 different lasers at 2 different wavelengths, and making sure that the resonance dips from both sensors don't overlap one another.

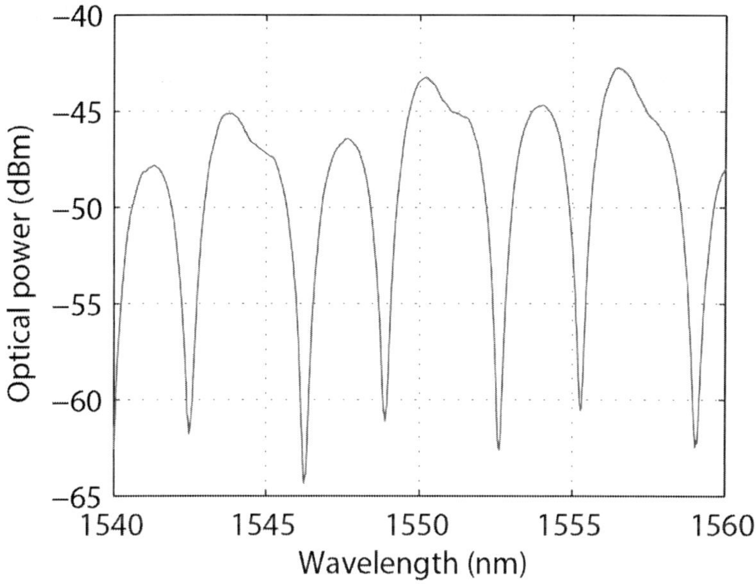

Figure 10. The push-pull SCOS has 2 lithium niobate crystals adhered to a single D fiber which causes a higher resonance dip frequency in the optical transmission spectrum.

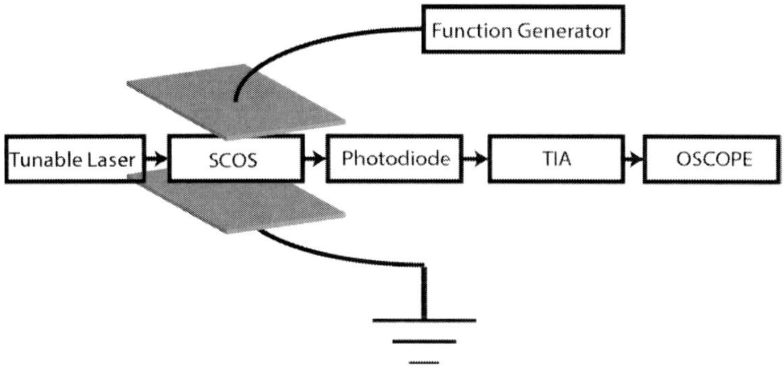

Figure 11. Push-pull SCOS fabrication interrogation setup.

Equation (1) states that the wavelength position of the resonance dip depends on the thickness of the lithium niobate crystal. Therefore to shift the wavelength of one of the crystals on the push-pull SCOS, a layer of silicon nitride was deposited using Plasma Enhanced Chemical Vapor Deposition (PECVD). Figure 8 illustrates this effect on a single SCOS crystal where its (red) resonance pattern after PECVD is shifted from its original (blue) resonance pattern before PECVD.

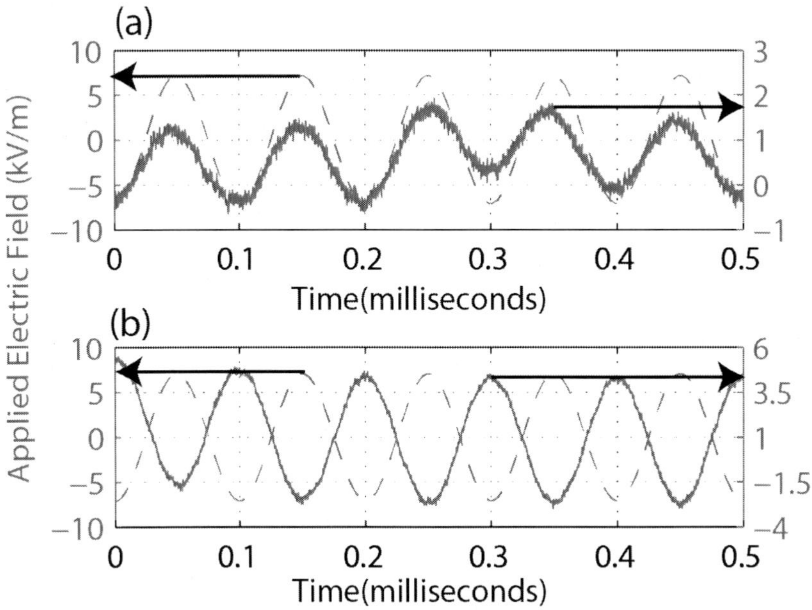

Figure 12. (a) Measured SCOS signal for uncoated crystal is in phase with applied electric field. (b) Measured SCOS signal is 180 degrees out of phase with applied electric field signal.

Figure 9 shows an image of the completed push-pull SCOS. The crystal that has undergone PECVD is notched in a corner to keep track of which surface has silicone nitride. Ideally the two crystals should be as close to each other as possible to ensure that the two SCOS measure the same noise. This concept is important for other differential sensor setups where it is important that the two sensors are localized to the extent that both sensors would measure the same noise signal. Figure 10 shows the final output spectrum of the push-pull SCOS where the spectrum is essentially a multiplication of the two spectra shown in Figure 8. 2 adjacent resonance edges are interrogated simultaneously. This way two measurements can be obtained from the push-pull SCOS that don't interfere with one another.

Figure 11 shows a setup used to illustrate this differential push-pull concept. A push-pull SCOS is mounted between two parallel plate electrodes and is attached to a tunable laser and a photodiode. The laser is launched at the rising edge of a SCOS resonance dip and the photodiode converts the modulated laser signal into an electrical current. The TIA then converts this electrical current into a voltage which is read by the Oscope. Figure 12 a) shows the actual applied electric field and the measured electric field when the laser is tuned to the second resonance edge from the left in Figure 10. As can be seen the measured electric field signal and the applied electric field signal are both perfectly in phase. Figure 12 b) shows the measured electric field signal after the laser has been aimed at the rising edge on the first resonance edge from the left in Figure 10. As can be seen, the two signals are now completely out of phase because the magnitude sign of the measured signal is opposite to that of the actual signal.

3.2. Electric Field Measurements

Figure 13 shows the setup for interrogating both SCOS on the push-pull simultaneously. Two lasers that are aimed at adjacent resonance edges are launched into a 50/50 fiber optic coupler/splitter which feeds both signals into the SCOS. A 50/50 splitter then feeds both signals into two different Lightwave2020 tunable mechanical fiber optic filters. Each filter is set to select a single wavelength from the push-pull SCOS, allowing 2 independent measurements to register on the Oscope.

Figure 14 shows the setup used to simulate local vibration noise. The SCOS is directly mounted on this structure so that it measures both strain noise and electric field simultaneously. The parallel plate electrode structure that applies the electric field across the SCOS is isolated from the impact to ensure that the electric field does not change during measurement. A weight is attached to a conductive swinging arm, and when the swinging arm hits the ground plate on the impact stage it creates a vibration local to the sensing element of the SCOS. The Oscope trigger channel is attached between the ground plate and the conductive swinging arm so that when an impact occurs the Oscope will be triggered via a short.

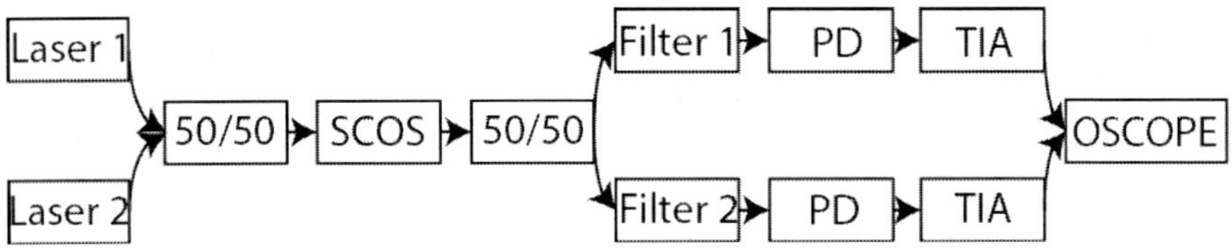

Figure 13. Push-pull SCOS interrogation setup.

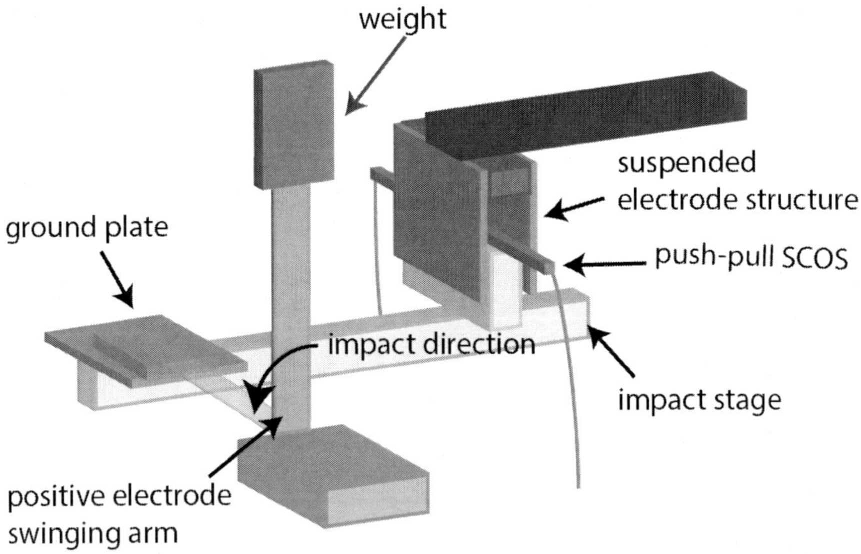

Figure 14. Impact stage setup for push-pull sensor. A positive electrode swinging arm impacts a ground plate attached to the impact stage which causes a short signal for the external trigger on the Oscilloscope.

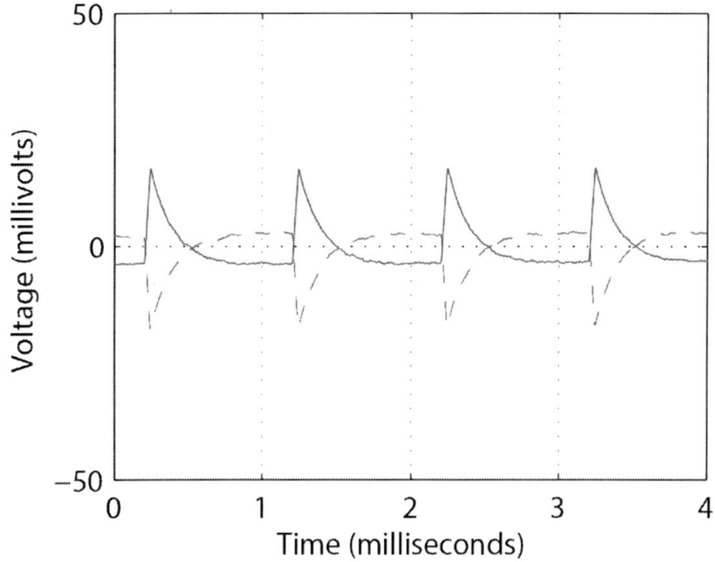

Figure 15. The push-pull SCOS obtains two opposite electric field measurements, one from each lithium niobate crystal.

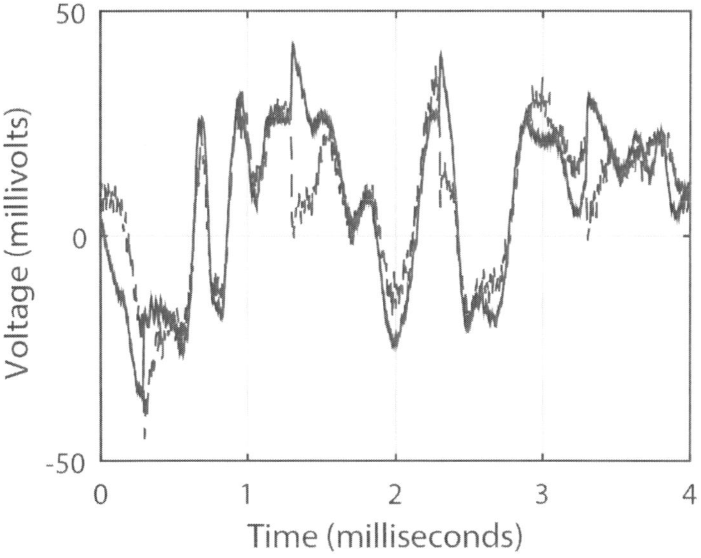

Figure 16. Vibration and electric field applied to the push-pull SCOS through two separate channels.

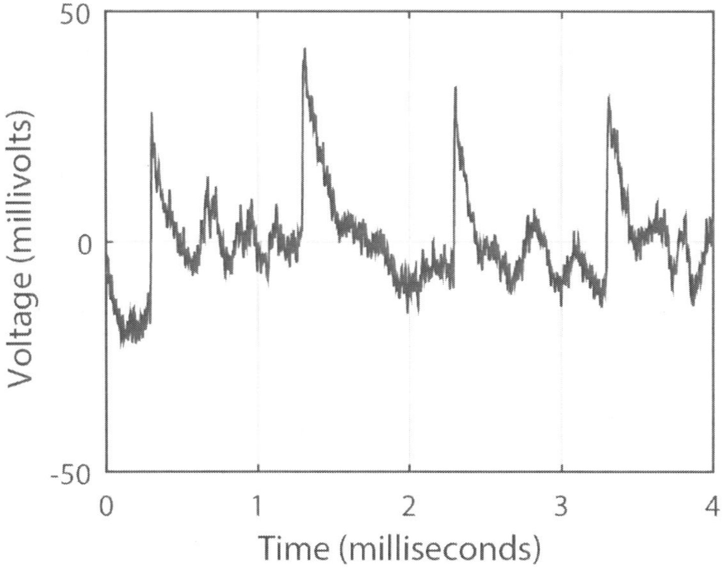

Figure 17. Subtraction of signals from Figure 16.

Figure 15 shows the two signals from the push-pull SCOS before impact, the same electric field as in Figure 5 is used. Figure 16 shows the signals from the push-pull SCOS with the same electric field applied, but during an impact. The localized strain noise tracks with the same magnitude sign, but the electric field signal tracks with opposite magnitude signs, allowing for a subtraction between the two signals to lead to the signal in Figure 17. The noise has been reduced and the electric field signal can be analyzed.

3.3. Summary

In summary, a differential method can be used to subtract out vibration induced strain noise on the SCOS. This is done by configuring 2 SCOS on the same optical fiber such that both sensors have the same response to strain and opposite response to electric field. Subtraction of the two simultaneous measurements allow for a reduction in the external noise and a recovery of the desired measurand.

4. Noise Isolation for Non-Localized Vibration Noise Reduction

In many systems the noise origin can be isolated and remedied physically rather than through signal processing. For example, in the previous section the sensor could have been mounted to foam or other shock absorbing material to reduce impact noise rather than needing to use a differential approach [16]. For this section the phase noise from the optical carrier is treated by phase modulation prior to its conversion to the amplitude domain. This prevents noise being manifest in the amplitude domain electric field measurement.

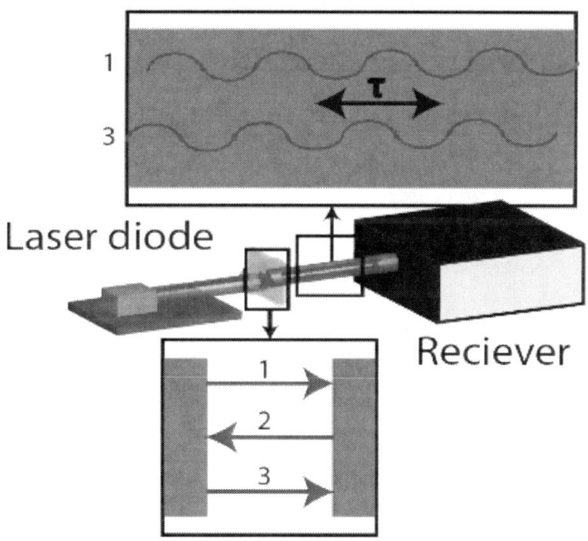

Figure 18. Multiple reflections at fiber connection points cause time shifted replicas of the optical carrier to propagate down the same fiber.

Although some noise types for a SCOS can be reduced with a differential setup, the SCOS itself has internal inherent noise which is made worse when placed in a vibrational environment. This is true for most sensors which have an internal noise floor. Figure 18 shows that the internal noise to the SCOS is phase induced intensity noise (PIIN), which is caused by double reflections at a fiber optic connection point. These double reflections form a fabry perot interferometer which allows two time shifted versions of the same optical carrier to propagate down the same fiber optic line. Therefore laser phase fluctuations or agitations to the phase of the two optical carriers by vibration/strain along the fiber is manifest in the intensity domain via interference.

4.1. Theory

For a derivation of the phase to intensity noise conversion process please see [23, 24]. This work only considers a single fabry perot interferometer in the system which is not necessarily true but proves to be simple and highly effective. The phase induced intensity I_n noise in a fiber optic network prior to phase modulation can be written as

$$I_n = 2\psi P_o \sqrt{d(t)d(t-\tau)}\cos\left(\omega_{op}\tau + \phi(t) - \phi(t-\tau)\right), \tag{2}$$

where P_o is the power from the optical source, ψ is the magnitude of the double reflection, $\phi(t)$ is the phase of the optical carrier, ω_{op} is the frequency of the optical carrier, τ is the time shift caused by the double reflection at a fiber optic connection point, and $d(t)$ is the measurand. Where

$$\Delta\phi(t) = \phi(t) - \phi(t-\tau) \tag{3}$$

is the phase noise term. The resulting generalized power spectral density of the noise is given by [24]

$$S_N(f) = 2\psi^2 P_o S_{sig} \otimes S_\phi, \tag{4}$$

where S_{sig} is the power spectral density of the signal, S_ϕ is the power spectral density of the phase noise, and $\otimes$ denotes convolution.

This interferometric noise is reduced by up conversion to a higher frequency band that enables the noise to be filtered out without removing the desired signal [25]. Figure 19 shows this is done by phase modulating the optical carrier via an external phase modulator driven by an RF generator. The phase modulator used in these experiments is a custom LN53S-FC from Thorlabs (Newton, NJ) with polarization maintaining fiber (PM) on the input and output for interrogation of the r_{33} coefficient of the lithium niobate crystal.

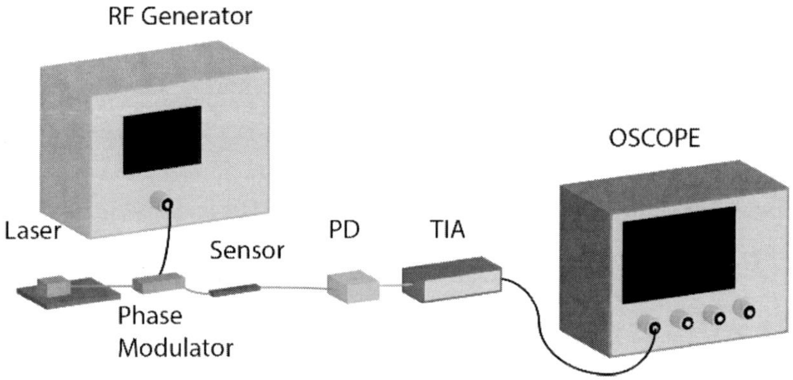

Figure 19. By phase modulating the optical carrier before feeding into the SCOS, interferometric noise due to random vibrations along the length of the fiber can be up-converted.

The phase modulation changes the noise intensity to [23]

$$I_n = 2\psi P_o \sqrt{d(t)d(t-\tau)} * \sum_{q=-\infty}^{\infty} J_q\left(-2k\sin\left(\frac{\omega_m \tau}{2}\right)\right) \\ * \cos\left(\omega_{op}\tau + \Delta\phi(t) + q\omega_m t - q\frac{\omega_m \tau}{2}\right), \quad (5)$$

where ω_m is the phase modulation frequency and k is the modulation index. Thus the phase modulator converts the vibration induced interferometric noise into a set of frequency bands centered at frequencies of $q\omega_m$. Figure 20 shows the first three frequency bands. Low-pass filtering the signal eliminates all the vibration induced interferometric noise terms except for the $q = 0$ term.

The remaining $q = 0$ term is the reduction in the interferometric noise and is called the noise reduction factor (NRF) and is given by

$$NRF = J_o^2\left(-2k\sin\left(\frac{\omega_m \tau}{2}\right)\right). \tag{6}$$

The time delay τ is determined by the sensor system components such as fiber length or the effective indices of the optical equipment used. Thus, there are two control parameters in Equation (6), the modulation frequency ω_m, and modulation index k.

Figure 21 shows the NRF simulated as a function of both k and $\omega_m\tau$. There are a repeating set of regions where the NRF is equal to 1 meaning that there is no reduction in the noise. These regions occur at

$$\omega_m\tau = 2\pi n, \tag{7}$$

where n is any integer. The greater the modulation index k the wider the range of frequencies with good NRF, therefore the maximum possible value of k from the RF generator was used in this work.

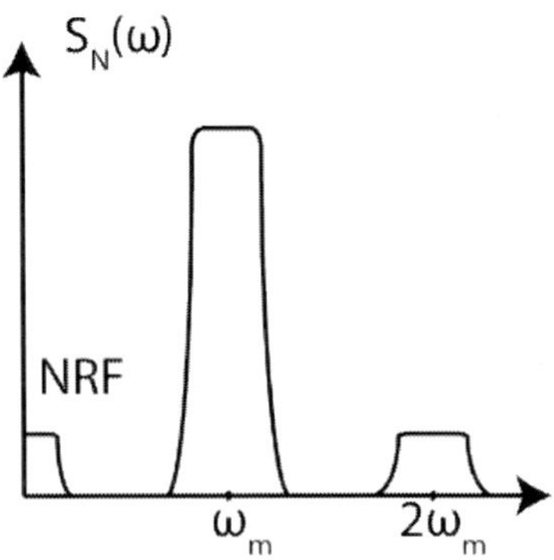

Figure 20. Distribution of the noise power among different harmonics after phase modulation.

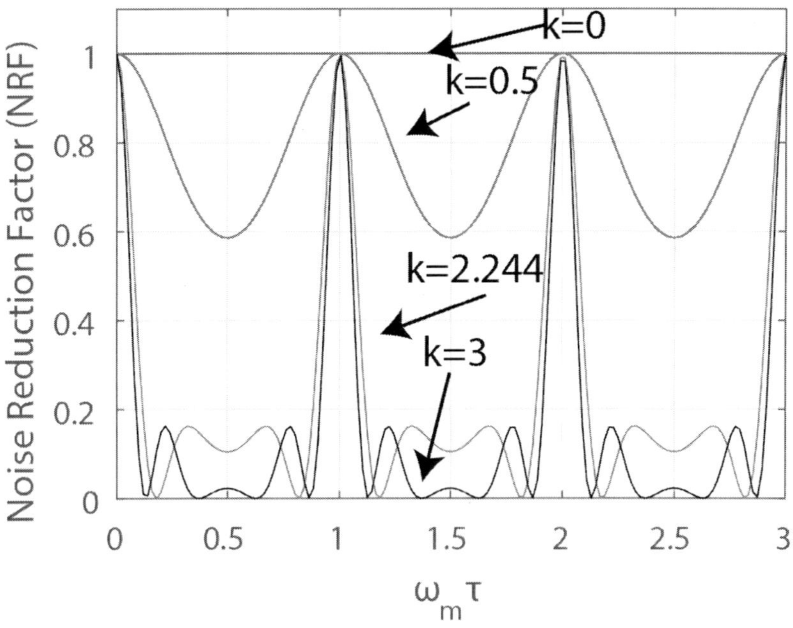

Figure 21. Noise reduction factor (NRF) as a function of the normalized modulation frequency, $\omega_m\tau$. There are certain modulation frequencies at which there is no noise reduction.

4.2. Measurements

The same electric field signal in Figure 5 is applied to the SCOS to produce the measurement shown in Figure 22 a). No vibration is applied to the fiber yet, but the interferometric noise is so bad in this setup that the periodic exponentials cannot be recognized at all. Figure 22 b) shows the Fourier transform of a). From this transform the dominant noise spike at 22 kHz can be identified. Figure 22 c) shows the electric field signal after phase modulating the optical carrier. The random phase fluctuations have been upconverted and the signal can be recognized. d) shows this upconversion effect where the spike at 22 kHz has been reduced dramatically. The periodic harmonics correspond to the Fourier transform of a periodic exponential signal.

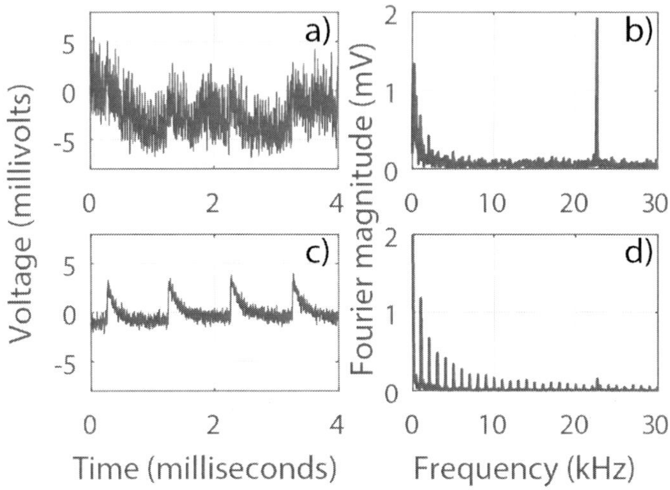

Figure 22. (a) Measurement of the low vibration system without phase modulation (b) Fourier transform of the low vibration system without phase modulation (c) Measurement of the low vibration system when phase modulation is applied (d) Fourier transform of the low vibration system with phase modulation.

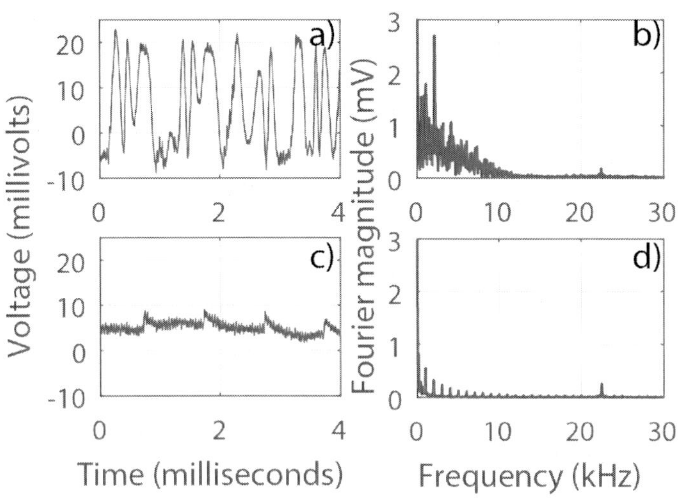

Figure 23. (a) Measurement of the slow vibration system without phase modulation. (b) Fourier transform of the slow vibration system without phase modulation. (c) Measurement of the slow vibration system when phase modulation is applied. (d) Fourier transform of the slow vibration system with phase modulation.

Vibrations along the length of the optical fiber sensing system can agitate the phase flucations even further leading to the noise seen in Figure 23 a). b) shows the Fourier transform for a) where the amount of interferometric noise energy has increased significantly due to vibrations. c) shows that single tone phase modulation upconverts most of the noise, allowing for the measured electric field signal to be discerned. This upconversion effect is particularly noticeable in the Fourier transform for c) shown in d) where most of the noise has been upconverted.

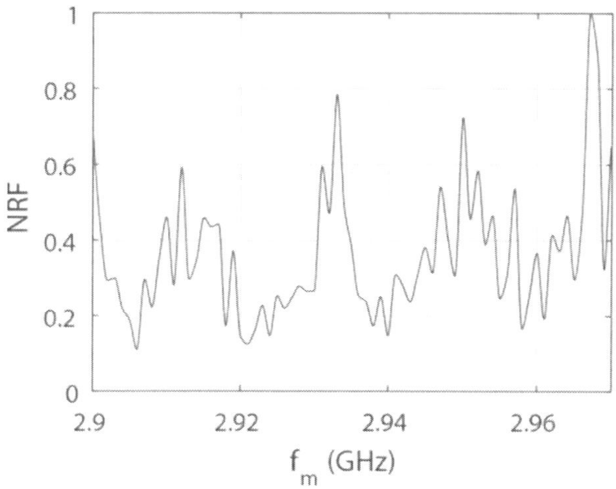

Figure 24. A suitable NRF can be found by sweeping f_m and choosing the best NRF.

In practice it is difficult to model the system such that the best NRF is found. Figure 24 shows that the NRF is found by sweeping ω_m across a range of frequencies to determine the best NRF. Figure 24 does not match up to Figure 21 exactly potentially due to multiple reflection points. However, single tone phase modulation was adequate for this application.

4.3. Summary

In summary, in some cases if it is possible to isolate the noise internal to the system from the measurand by domains or other means, then it is possible to reduce that noise physically rather than through signal processing. In this subsection, noise that is manifest in the amplitude domain from the phase domain is first upconverted in the phase domain. This noise is then filtered by the electronics in the system, allowing for the amplitude domain signal to be recovered.

5. Simultaneous Noise Reduction for Optical Sensing of Electric fields in Harsh Environments

Two noise reduction techniques have been illustrated. One in reducing the noise in the phase domain such that the signal in the amplitude domain remains unaffected. The second in reducing localized vibration noise on a sensor system via a differential setup such that the measurand can be properly analyzed. However, in many situations there are multiple noise sources present which contribute to a harsh environment, as a result it is necessary for sensor systems to be able to compensate for multiple noise types simultaneously to allow for sensing in harsh environments. This section combines the noise reduction methods from the previous two sections to allow for optical sensing of electric fields in harsh environments with a SCOS.

5.1. Harsh Environment Setup

Figure 25 shows that the push-pull SCOS and phase modulation technique are combined to take out different noise types simultaneously. The interrogation setup is essentially the same as for a push-pull SCOS, except both optical carriers are now phase modulated via phase modulators PM1 and PM2 before being fed into the system. RF1 and RF2 are the RF generators driving PM1 and PM2. The different noise types reduced in this section are phase induced intensity noise, localized vibration noise, and RF noise. The harsh environment used in this section is shown in Figure 26. It is a large wooden swinging arm with medical tubing attached between the free end and the platform that it is mounted on.

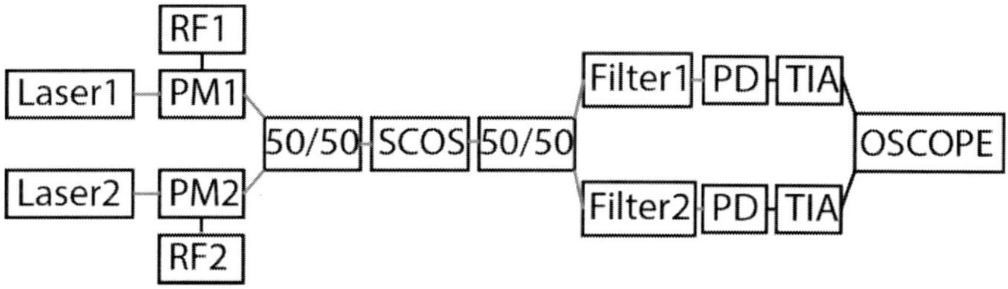

Figure 25. The measurement setup for optical sensing for electric fields in harsh environments involves independently phase modulating each optical carrier of the push-pull, and then using the push-pull configuration [22] to subtract out any local strain noise on the sensor.

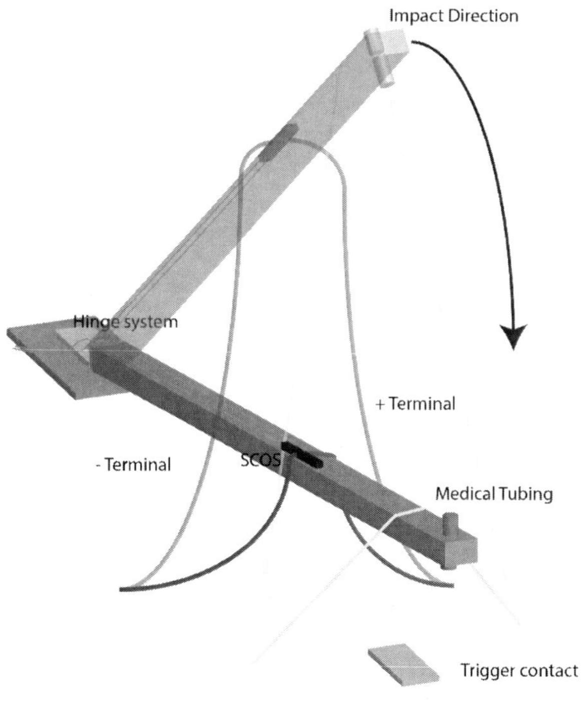

Figure 26. Swinging arm apparatus used to apply random vibration noise to the sensor and sections of fiber attached to the sensor. The data acquisition system is triggered via a trigger contact on the large swinging arm.

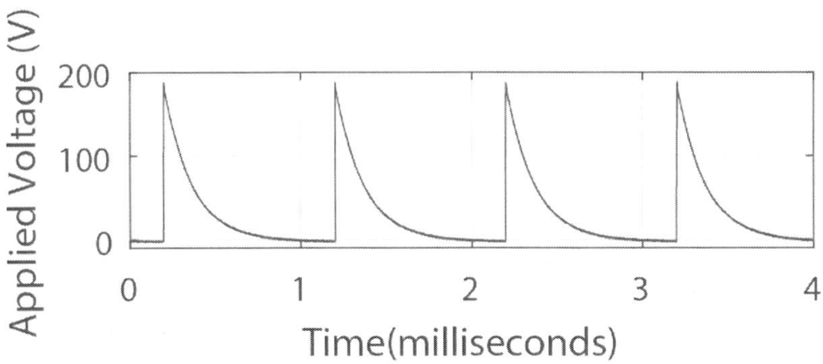

Figure 27. The voltage signal applied to the electrodes placed on the sides of the SCOS.

The SCOS is packaged in a plastic box and attached to the large swinging arm. There are electrodes on either side of the packaging box where a voltage is applied to generate an electric field. Figure 27 shows the voltage applied to the electrodes, which are spaced by *18.5 mm*. The resulting applied electric field is a periodic set of exponential pulses with amplitude of *10 kV/m*. The Oscilloscope is triggered on impact by contacting a conductor on the free end of the arm with a ground plate. This setup induces vibrations along the entire length of the fiber rather than just locally at the SCOS location which agitates interferometric noise and induces a large amount of strain on the SCOS. Therefore, in this setup the push-pull SCOS would not work without phase modulation and vice versa.

5.2. Harsh Environment Noise Reduction

Figure 28 a) shows the desired electric field measurement from the push-pull SCOS. b) shows that the field is completely unrecognizable after the large swinging arm impacts the trigger contact. The two signals have minimal correlation due to the interferometric noise. c) shows that the two signals exhibit much better correlation after the interferometric noise is upconverted and can be subtracted to produce the signal shown in d). It is uncertain whether the leftover noise is due to strain that cannot be subtracted out or due to the electrodes are shaking during impact.

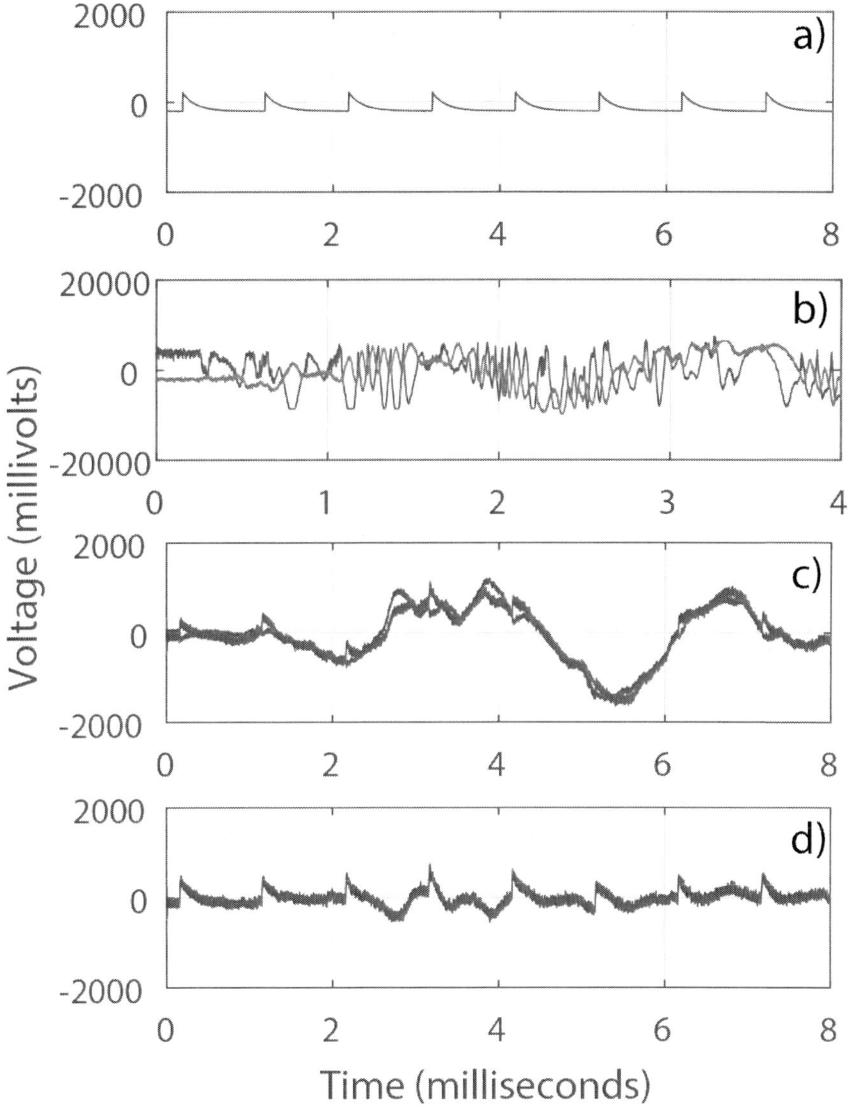

Figure 28. (a) Electric field applied to the push-pull SCOS. (b) Measured electric field from harsh environment on two channels without phase modulation (c) Measured electric field on two channels with phase modulation. (d) Subtraction of the two signals in (c).

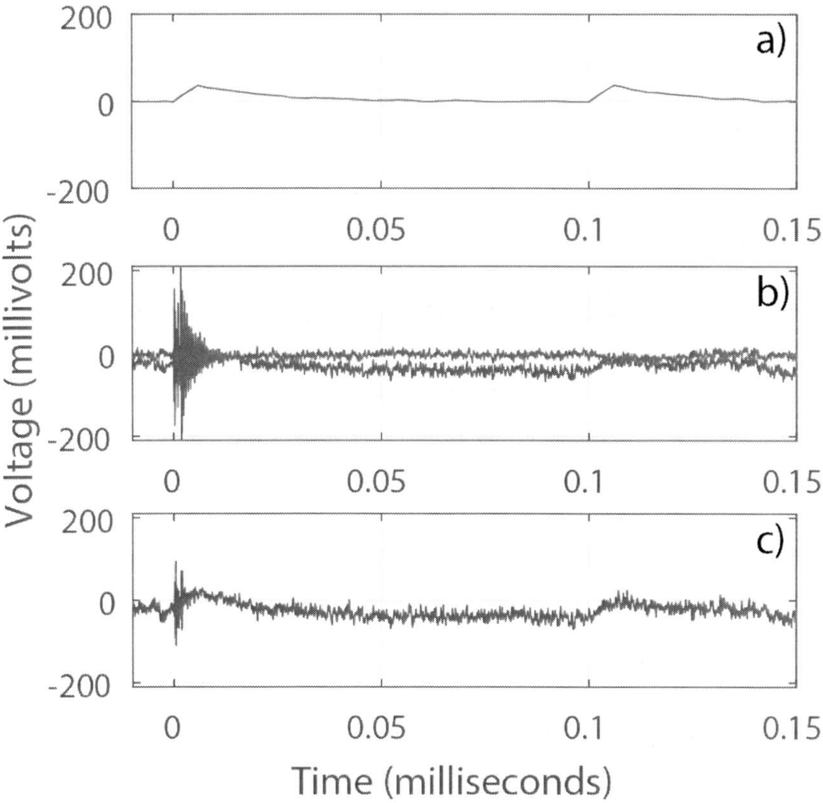

Figure 29. (a) Electric field applied to the SCOS. (b) Measured signal from the two SCOS sensing elements. (c) Subtraction of the two SCOS signals.

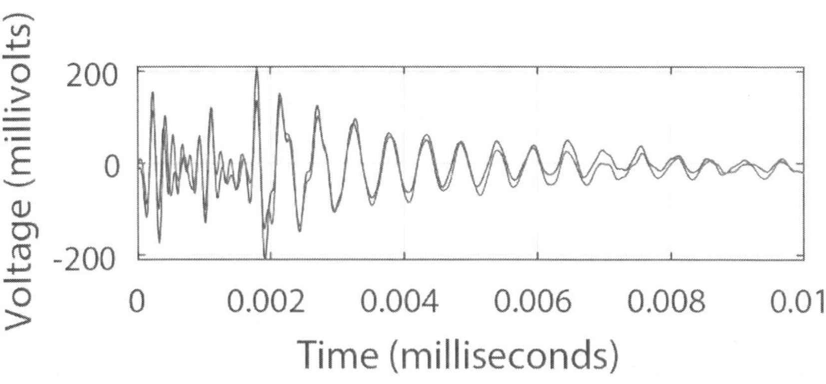

Figure 30. Zoomed-in image of Figure 29 (b), the RF noise on both channels track, allows for a push-pull subtraction of the RF noise.

There are a variety of systems under test that produce RF noise [17, 18]. Fortunately, the push-pull SCOS measures the electric field in opposite magnitude signs, while the electronics pick up RF noise with the same magnitude sign. As a result, the push-pull can also be used to subtract out RF noise. Figure 29 a) shows that to observe RF noise the frequency of the periodic exponentials is increased from 1kHz to 10 kHz such that the time window on the Oscope has been shrunk. b) shows the measured signal from the push-pull on both channels when the large swinging arm impacts the contact trigger. The interferometric noise has been upconverted and the strain noise is slower than this 10 kHz period so it is irrelevant and not noticeable in the measurement. However, the RF noise at the trigger time of the signal distorts the signal and does not allow for an analysis of crucial elements such as rise time. c) shows the deduced signal after subtracting the two signals in b). The RF noise has been reduced dramatically and the signal can be better analyzed. Figure 30 shows the RF noise tracking on both channels of the push-pull SCOS interrogation system. Discrepencies could be slight differences in the equipment manufacturing on either channel or potentially residual interferometric noise.

6. Summary and Application to Other Sensors

Fiber optic sensors in harsh environments must be able to overcome all the noise types simultaneously to obtain a proper measurement. This work has illustrated that by independently single tone phase modulating each optical carrier channel of the push-pull SCOS, it is possible to eliminate non-local vibration noise which manifests itself as interferometric noise. This allows for effective use of the push-pull SCOS configuration, which drastically reduces local vibration noise to the sensing element and RF noise, which manifests itself as damped oscillations for shorter time periods.

The setup used in this work can reduce all noise sources individually or simultaneously. This allows for the SCOS to measure electric fields in a harsh environment. These principles can also be applied to other sensors to allow for reduction of multiple noise sources simultaneously. The bottom subsection illustrates how to eliminate phase induced intensity noise in fiber Bragg gratings.

6.1. Phase Induced Intensity Noise Reduction in Fiber Bragg Gratings

Figure 31 shows that a fiber Bragg grating (FBG) is a section of optical fiber with a periodic variation in the refractive index of the core. The FBG reflects a specific wavelength as given by

$$\lambda_B = 2n_e\Lambda, \tag{8}$$

where λ_B is the reflected wavelength called the Bragg wavelength, n_e is the effective refractive index of the optical fiber mode and Λ is the grating period.

When the FBG is strained it causes a change in both the grating period Λ and the effective index of refraction of the fiber mode n_e resulting in a shift of the transmission spectrum. For an FBG written in a standard optical fiber the shift is *1.2 pm/$\mu\varepsilon$* [26]. Very small strain can be measured by tuning a laser to have a wavelength that lies on the edge of the transmission dip. Small shifts in the spectrum result in changes in the transmitted optical power. This type of FBG interrogation is called FBG edge detection [27, 28]. The FBG edge detection interrogation system is like the SCOS interrogation system and consists of coupling a tunable laser into the optical fiber with the FBG. The optical power transmitted through the FBG is then converted into a current by a photodiode and then into a voltage using a transimpedance amplifier. The voltage signal is then captured using an oscilloscope.

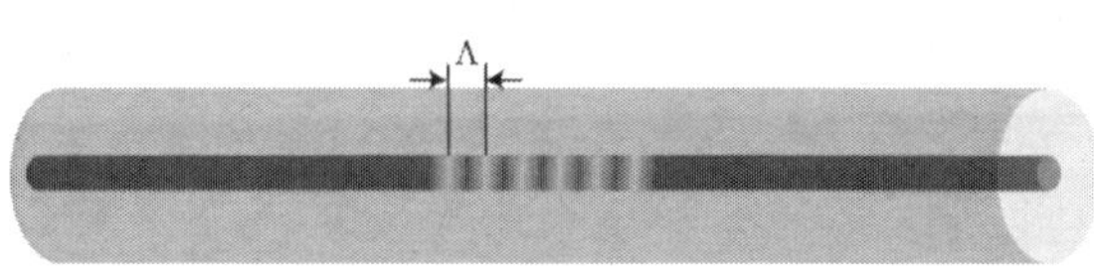

Figure 31. Cross sectional figure of a fiber Bragg grating.

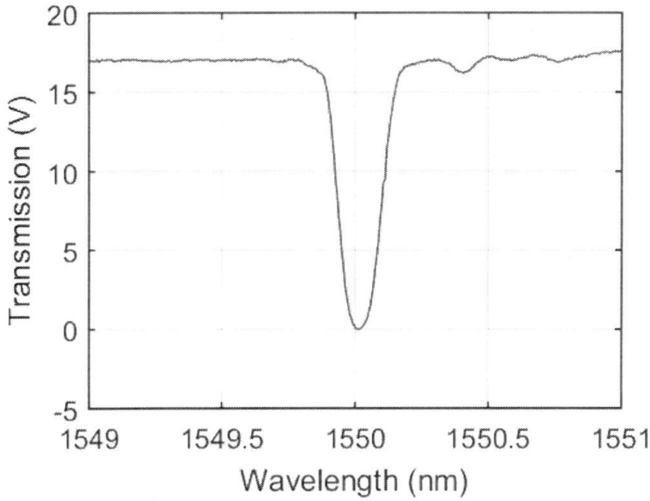

Figure 32. The measured FBG spectrum.

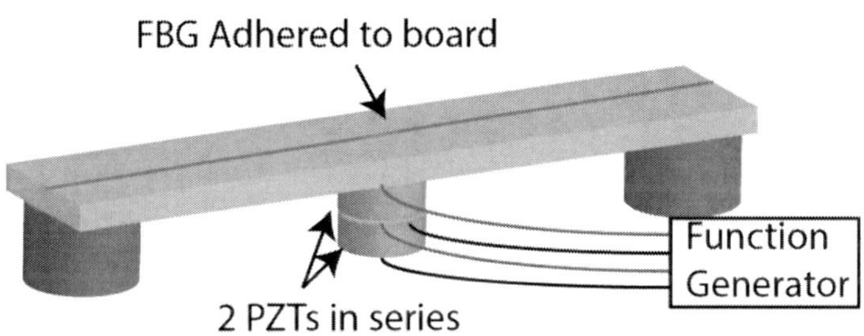

Figure 33. The FBG measurement system.

Figure 33 shows that the FBG measurement setup consists of an FBG attached to a foam board using Devcon 5-minute epoxy. The board is attached to solid platforms at both ends to prevent excessive movement. A piezoelectric transducer (PZT) is attached using 3M industrial double-sided tape to the plastic board. The PZT is driven with a sinusoidal signal at a frequency of 2 kHz.

Figure 34 (top) shows the FBG measurements before external phase modulation of the optical source. Since the FBG measurement uses different lengths of optical fiber than the SCOS measurement setup it has a different value for τ. The different value for τ required that the phase modulation frequency be adjusted to attain a good NRF. Figure 34 shows that with the phase modulation the noise is reduced by a factor of 25.

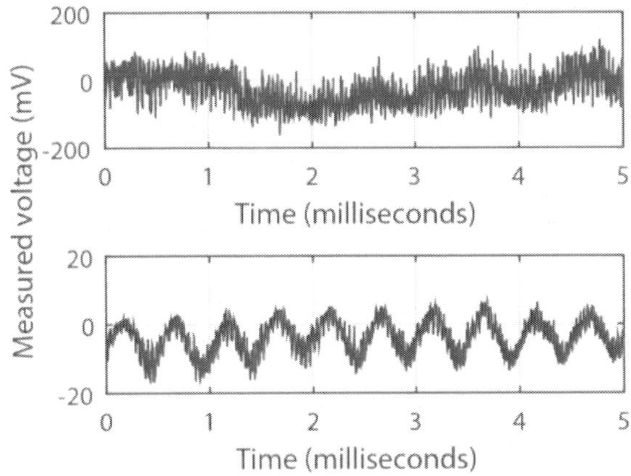

Figure 34. FBG measurement (top) before phase modulation, and (bottom) after phase modulation.

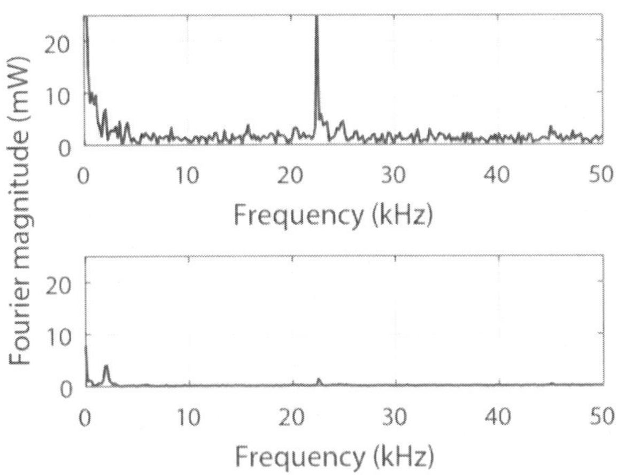

Figure 35. Fourier transform of FBG measurements in Figure 34 (top) before phase modulation, and (bottom) after phase modulation.

Figure 35 shows the Fourier transform of the FBG measurement. Like the SCOS measurement, there is an interferometric noise spike at 22 kHz caused by the laser phase noise. With the phase modulator the interferometric noise has been reduced enough to enable the 2 kHz strain signal to be measured. With the reduction in the interferometric noise the 2 kHz strain signal is measured to have an amplitude of 14 mV, which corresponds to a strain with an amplitude of 0.1 με.

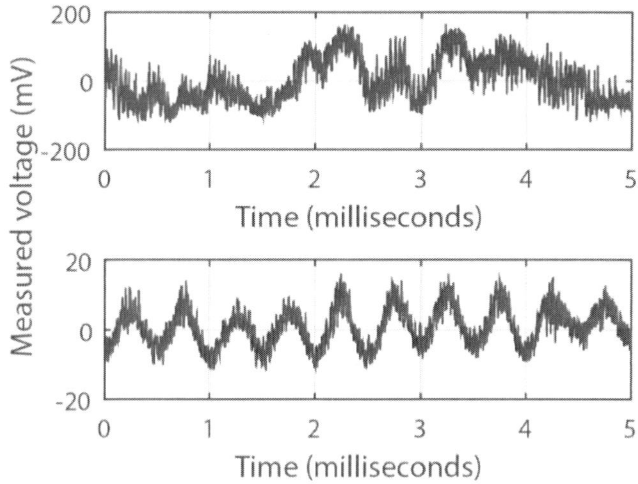

Figure 36. (top) measurement with FBG in a dynamic environment without phase modulation and (bottom) with phase modulation.

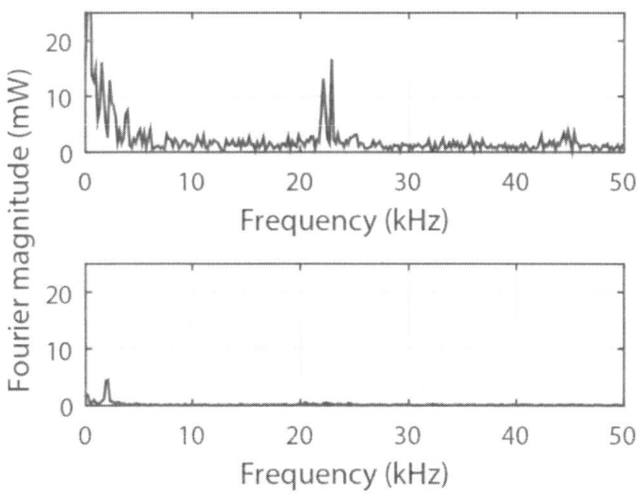

Figure 37. Fourier transform of FBG measurements in Figure 34 (top) before phase modulation, and (bottom) after phase modulation.

Figure 36 shows the FBG measurement when a manual vibration is applied to the FBG output fiber. The noise is reduced by a factor of 26. Figure 37 shows that the noise has been reduced dramatically and the Fourier spectrum essentially only contains the 2 kHz sinusoid signal.

6.2. Summary

As T&E requirements become more demanding on sensors, it is important that the sensors can withstand harsh environments to the extent that the signal being measured can be properly discerned from the noise. There are many different methods for noise reduction in sensors, this chapter illustrated two primary methods: A differential sensor setup, and noise upconverstion in the phase domain to prevent noise manifestation in the amplitude domain. The work in this chapter was illustrated on fiber optic sensors but is applicable to other sensor systems as well.

References

[1] D. S. Ballantine, S J. Martin, A J. Ricco, G C. Frye, H Wohltjen, R. M. White, and E. T. Zellers. *Acoustic wave sensors Theory, design, and physico-chemical applications.* San Diego: Academic Press, 1997.

[2] C. M. Jha, *Thermal Sensors Principles and Applications for Semiconductor Industries*. New York, Springer New York, 2015.

[3] K. Poletkin, and L. A. Francis. *Magnetic Sensors and Devices: Technologies and Applications.* Taylor & Francis Group, LLC, 2018.

[4] E. Garcia-Breijo, B. G. Perez, and P. Cosseddu. Title: *Organic Sensors: Materials and Applications*. London, The Institution of Engineering and Technology, 2016.

[5] J. Chen, S. He, and Y. Sun. Title: *Rechargeable Sensor Networks: Technology, Theory, and Application - Introducing Energy Harvesting To Sensor Networks*. Singapore, World Scientific Publishing Co. Pte. Ltd., 2014.

[6] S. Kang, A. Paul, and G. Lee. Reduction of mixed noise from wearable sensors in human-motion estimation. *Computers & Electrical Engineering*, vol. 61, July 2017, pp. 287-96.

[7] Boukhayma, A. Peizerat, and C. Enz. Temporal Readout Noise Analysis and Reduction Techniques for Low-Light CMOS Image Sensors. *IEEE Transactions on Electron Devices*, vol. 63, no. 1, 1 Jan. 2016, pp. 72-78.

[8] Fang, J. Ma, M. Yao, W. Di, and D. Lin, H. Xu, W. Wang, and H. Luo. Equivalent magnetic noise reduction at high frequency range due to polarized direction optimization in Terfenol-D/Pb(Mg1/3Nb2/3) O3-PbTiO3 magnetoelectric laminate sensors. *Journal of Magnetism and Magnetic Materials*, vol. 423, 1 Feb. 2017, pp. 106-10.

[9] Y. Wang, J. Li, and D. Viehland. A differential magnetoelectric heterostructure: Internal noise reduction and external noise cancellation. *Journal of Applied Physics*, vol. 118, no. 21, 2 Dec. 2015, pp. 214103-1--5.

[10] M. Lassen, A. Brusch, D. Balslev-Harder, and Jan C. Petersen. Phase-sensitive noise suppression in a photoacoustic sensor based on acoustic circular membrane modes. *Applied Optics*, vol. 54, no. 13, May 2015, pp. D38-42.

[11] S. Chadderdon, Application Improvements of Slab-Coupled Optical Fiber Sensors (2014). *All Theses and Dissertations*. 3976. https://scholarsarchive.byu.edu/etd/3976.

[12] N. Stan, F. Seng, L. Shumway, R. King, R. Selfridge, and S. Schultz, "High electric field measurement using slab-coupled optical sensors," *Appl. Opt.* 55, 603-610 (2016).

[13] C. Y Lin, A. X. Wang, B. S. Lee, X. Y. Zhang, and R. T. Chen, High dynamic range electric field sensor for electromagnetic pulse detection, *Opt. Express* 19, 17372-17377 (2011).

[14] E. K. Johnson, J. M. Kvavle, R. H. Selfridge, S. M. Schultz, R. Forber, W. Wang, and D.Y. Zang, Electric field sensing with a hybrid polymer/glass fiber, *Appl. Opt.* 46, 6953-6958 (2007).

[15] F. Seng, N. Stan, R. King, R. Worthen, L. Shumway, R. Selfridge, and S. Schultz, Optical Sensing of Electrical Fields in Harsh Environ-

ments, in *Optical Fiber Communication Conference*, OSA Technical Digest (online) (Optical Society of America, 2016), paper M2D.3.

[16] J. R. Noren, Electric Field Sensing in a Railgun Using Slab Coupled Optical Fiber Sensors (2012). *All Theses and Dissertations.* Paper 3482. http://scholarsarchive.byu.edu/etd/3482.

[17] R. B. Hoffman, T. L. Haran, and S. R. Lane. Diagnostic capabilities for electromagnetic railguns. *Electromagnetic Launch Technology (EML), 2012 16th International Symposium on* (2012).

[18] B. Fryskowski. Development of vacuum-tube-based voltage and current probes for automotive ignition systems. *Proceedings of the Institution of Mechanical Engineers, Part D: Journal of Automobile Engineering* 229.8 (2015): 958-68.

[19] S. Chadderdon, L. Woodard, D. Perry, R. H. Selfridge, and S. M. Schultz, Single tunable laser interrogation of slab-coupled optical sensors through resonance tuning, *Appl. Opt.* 52, 2682-2687 (2013).

[20] R. Gibson, R. Selfridge, S. Schultz, W. Wang, and R. Forber. Electro-optic sensor from high Q resonance between optical D-fiber and slab waveguide. *Applied Optics* 47. 13 (2008): 3505-12.

[21] F. Seng, N. Stan, R. Selfridge, and S. Schultz. Non-Intrusive High Voltage Measurement Inside a Coaxial Cable Using a Slab-Coupled Optical Sensor (SCOS). *Journal of Directed Energy*, vol. 5, no. 4, 2016, pp. 335-45.

[22] F. Seng, N. Stan, C. Josephson, R. King, L. Shumway, R. Selfridge, and S. Schultz, Push–pull slab coupled optical sensor for measuring electric fields in a vibrational environment, *Appl. Opt.* 54, 5203-5209 (2015).

[23] F. Seng, N. Stan, R. King, C. Josephson, and L. Shumway. Optical Sensing of Electric Fields in Harsh Environments. *Journal of Lightwave Technology*, vol. 35, no. 4, 15 Feb. 2017, pp. 669-76.

[24] P. K. Pepeljugoski, and K. Y. Lau. Interferometric Noise Reduction in Fiber-Optic Links by Superposition of High Frequency Modulation. *Journal of Lightwave Technology* 10.7 (1992).

[25] E. H. Chan, Suppression of phase-induced intensity noise in fibre optic delay line signal processors using an optical phase modulation technique. *Optics Express* 18. 21 (2010).

[26] F. Seng, D. Hackney, T. Goode, L. Shumway, A. Hammond, G. Shoemaker, M. Pankow, K. Peters, and S. Schultz, (2016). Split Hopkinson bar measurement using high-speed full spectrum fiber Bragg grating interrogation. *Applied Optics*, 55(25), 7179-7185.

[27] R. Romero, O. Frazao, P. V. S. Marques, H. M. Salgado, and J. L. Santos. Fibre Bragg grating interrogation technique based on a chirped grating written in an erbium-doped fibre. *Measurement Science & Technology* 14. 11 (2003): 1993-97.

[28] U. Tiwari, K. Thyagarajan, M. R. Shenoy, and S. C. Jain. EDF-Based Edge-Filter Interrogation Scheme for FBG Sensors. *Sensors Journal, IEEE* 13. 4 (2012): 1315-19.

About the Authors

Frederick Seng received the Bachelor's degree in electrical engineering in 2014 from Brigham Young University, Provo, UT, US, where he is currently working toward a PhD degree in electrical engineering with an emphasis on fiber optic sensors for electric field detection in harsh environments and dynamic shape sensing. He has interned at Northrop Grumman at Hill Air force base as well as two different startup companies.

Nikola Stan has earned a PhD degree in electrical engineering from Brigham Young University, Provo, Utah in 2018 and a BSc in electrical engineering from University of Belgrade, Serbia in 2010. His research topics include measuring high electric field and voltage using slab-coupled optical fiber sensors and increasing dynamic range of a fiber Bragg-grating interrogator using a feedback control loop. Most recent publications include "Non-perturbing voltage measurement in a coaxial cable with slab-coupled optical sensors," published in Applied Optics Vol. 56, Issue 24 in 2017 and

"High electric field measurement using slab coupled optical sensors," published in Applied Optics Vol. 55, Issue 3 in 2016.

Rex King earned a bachelors and master's degree from Brigham Young University, Provo, Utah in 2013 and 2015. He focused on electro optic sensors and published "Slab–coupled optical sensor fabrication using side-polished Panda fibers" in applied optics. He currently works as a systems engineer in Salt Lake City.

Stephen M. Schultz earned both a BS and MS degree from Brigham Young University, Provo, UT, US, in 1992 and 1994, respectively, and the Ph.D. degree from the Georgia Institute of Technology, Atlanta, GA, US, in 1999, all in electrical engineering. He worked at Raytheon Missile Systems from 1999 to 2001. He has taught at Brigham Young University since 2002 and is currently a Full Professor. He has authored or coauthored more than 100 publications and holds 10 patents. His research interests include the area of optical fiber devices with an emphasis on optical fiber-based sensors.

In: Noise Reduction
Editor: Marc Farrow
ISBN: 978-1-53613-541-1

Chapter 4

NOISE REDUCTION IN ULTRASONIC NON-DESTRUCTIVE EVALUATION (NDE) AND IMAGING

***Minghui Li*[1,*] *and Gordon Hayward*[2]**
[1]School of Engineering, University of Glasgow, Glasgow, UK
[2]Alba Ultrasound, Todd Campus, West of Scotland Science Park, Glasgow, UK

ABSTRACT

Reliable and robust defect detection is a challenging yet essential problem in ultrasonic non-destructive evaluation (NDE). The flaw echoes are usually contaminated by high-level time-invariant grain noise originating from the material microstructures. This phenomenon becomes even worse when inspecting coarse-grained materials like austenitic stainless steel, alloy, carbon-reinforced composite and concrete, which however, form some of the most important and commonly used industrial materials. This problem has attracted great attentions from the NDE

* Corresponding Author Email: minghui.li@ieee.org.

community and a wide variety of techniques have been investigated to enhance the signal-to-noise ratio (SNR) and flaw detection. In this chapter, the signal processing approaches that exploit the spatial and temporal-spectral characteristics of the flaw echo and noise for enhanced defect detection and noise reduction in ultrasonic NDE are presented. The first category of techniques exploits the spatial diversity created by the transducer array, and utilises spatial filtering to discriminate the flaw echo from the clutter noise. Particularly, minimum variance beamforming seeks to minimize the power of the array output while maintaining a unit gain on the focal point, and weighs the observations with data-dependent apodization. The second category of approaches investigates the time-frequency analysis and statistical properties of flaw echoes and grain noise. In particular, the advanced matched filter is designed to match unknown flaw echoes rather than the transmitted signal, and utilise an optimization paradigm to tweak and search the best filter parameters. Results from simulation studies and experiments on the real industrial samples are presented to evaluate the techniques, and demonstrate their robustness and advantages over the classical methods.

Keywords: non-destructive evaluation, ultrasound, matched filtering, adaptive beamforming

1. INTRODUCTION

Reliable and robust flaw detection in the presence of grain noise is a challenging yet essential problem in ultrasonic non-destructive evaluation (NDE), which has attracted significant attention from the NDE community in recent decades [1-17]. A variety of coarse-grained materials like alloy and austenitic stainless steel offer attractive properties like high-temperature strength or excellent resistance to corrosive environment, and thus are widely used to build components like ducting, combustion cans, and transition liners in a range of key industrial sectors such as energy, oil and gas, nuclear, and aerospace. However, those materials composed of coarse-grained structures may pose significant challenges for ultrasonic pulse-echo inspection techniques, because the flaw echoes are usually contaminated by significant grain noise originating from the material microstructures, furthermore, the grain noise is time-invariant, demonstrating similar spectral characteristics, and correlated with the flaw echoes.

A wide variety of techniques have been investigated to suppress grain noise and enhance flaw detection through exploiting the key differences between defect echoes and grain noise. Since the defect (or a legitimate reflector) and grain boundaries often have different sizes, their scattering properties and responses have different spectral characteristics [3]. These differences have motivated temporal-spectral processing techniques given that the excitation signals are usually broadband, such as Weiner filtering [4, 5], split-spectrum processing [6-8], fragment recognition classifier [9], sub-spectrum phase coherence factoring [10], spectral distribution similarity analysis [11], and matched filtering [12, 13]. The defect echo typically has a coherent structure with energy mainly scattered from a single spatial point, while grain noise are spatially distributed throughout the insonified resolution cell. These differences have motivated spatial diversity and processing techniques, such as spatial beamforming [2, 14], coherence factoring [15], and spatially averaged sub-array correlation imaging [16, 17].

This chapter focuses on advanced matched filtering and spatial beamforming techniques for noise reduction in ultrasonic NDE. The signal matching concept has been extensively used in the detection of signals of known form in stationary noise in applications like radar and sonar [18], and more recently in the field of ultrasonic NDE [19, 20]. If the signal waveform and noise statistics are exactly known *a priori*, the matched filter is optimal in terms of the signal-to-noise ratio (SNR) improvement; this is unfortunately not the case in NDE. In this chapter, the matched filter design is investigated and extended into the scenario of inspection of coarse-grained materials. The filter is tuned to match the unknown defect echoes (rather than the transmitted signal) which are approximated by the superposition of multiple transmitted signals with different phase shift, time delay and amplitude gain that simulate reflections from an unknown spatially extended defect. An optimization paradigm is employed to search for the optimal parameters in the filter response in order to maximize the SNR gain over a set of training signals.

In addition to matched filtering, this chapter presents a spatial filtering approach which utilises adaptive beamforming in reception to suppress grain noise and enhance defect echoes, with the assumption that they originate from different spatial locations. Adaptive beamforming applies a delay-weight-and-sum operation to the received signals across the probe aperture, where the adaptive weights are computed from the statistical analysis of the data samples based on the minimum variance theory. The ways of increasing robustness in NDE scenarios are considered.

The chapter is organised as follows. Section 2 presents the data model for the defect and grain echoes in ultrasonic NDE, and discusses the design and implementation of the optimised matched filter. Section 3 formulates the problem of spatial beamforming, presents the minimum variance adaptive beamformer, and discusses the techniques to enhance its robustness in the presence of mismatches or errors like wrong assumptions of acoustic velocity or phase aberrations. Section 4 presents experiments on real industrial samples and simulation studies to evaluate the performances of the adaptive beamformer and the optimised matched filter for noise reduction and defect detection against the classical methods. The chapter is then concluded.

2. Optimal Matched Filter Design in Temporal-Spectral Domain

2.1. Data Model in Ultrasonic NDE

This section describes the ultrasonic data model, which is used for the optimal matched filter design. If there is a single point reflector in a homogeneous and lossless propagation media, the return signal $x(t)$ at the receive (rx) element is equal to the signal $s_0(t)$ transmitted from the transmit (tx) element except for a time delay t_0 and scaling factor A,

$$x(t) = As_0(t - t_0). \quad (1)$$

The propagation distance from the tx element to the point reflector and back to the rx element can be calculated from the coordinates of the transmit and receive elements and the reflector. And the propagation time, or time delay t_0, is determined by dividing the propagation distance by the longitudinal velocity of sound in the media. The amplitude scaling factor A is employed to address the effects of element directivity and beam spread, as well as the effect of the divergence of the waves from the tx element and from the point reflector. The above-mentioned point reflector signal model is straightforward to follow, and has been used in a variety of literature on radar [21, 22], communications [23, 24], and ultrasonic imaging [25] as a standard data model.

However, in most practical ultrasonic NDE applications, the defects cannot be simply represented by single or well isolated point reflectors; instead most defects are spatially extended and distributed with a spatial shape or profile. A straightforward yet reasonable extension to the above data model for a distributed defect is that the defect consists of a number of point-like reflectors, each with given scaling and characteristics due to the effect of element directivity, beam spread, and spatial position that determines the propagation time. For a spatially extended defect, the return signal $x(t)$ received at the rx element is then given by a sum of J delayed and scaled versions of the transmitted signal $s_0(t)$,

$$x(t) = \sum_{i=1}^{J} A_i s_0(t - t_i) + n(t), \tag{2}$$

where $s_0(t)$ is the excitation signal taking into account of the pulser output waveform and the frequency response of the transducer, J is the number of point-like reflectors under consideration for this particular extended defect, A_i and t_i are the amplitude factor and the propagation time delay corresponding to the ith point reflector with certain position and reflection characteristics, and $n(t)$ is the additive noise due to the transceiver internal circuit and the environment noise in the media and couplant, which is assumed to be uncorrelated with the defect echoes.

2.2. Optimal Matched Filter Design

As shown in the data model (2), the return echo originating from a defect is contaminated by noise, which may consist of the electronic noise in the transmitter/receiver channel circuits, the large amount of reflections from the microstructure grain boundaries within the test sample, the coupling variation between the transducer and the sample, and etc.; and the noise could be significant with regard to the signal of interest originating from the defect, and thus results in a poor SNR. Matched filtering is to pass the echoes at the receiver through a filter with a particular response to suppress the noise and enhance the SNR. Assume $x(t)$ is a returned echo from a distributed defect received by a transducer, which is processed by a matched filter with an impulse response $h(t)$, the output signal $y(t)$ is given by

$$y(t)=x(t)*h(t), \tag{3}$$

where * stands for the convolution operation. $x(t)$ is composed of the signal component which is determined by the defect profile and characteristics (shape and reflection factor) as well as the additive noise, as illustrated in the data model (2). Based on the theory of matched filtering, if the filter response $h(t)$ is exactly matched to the echo reflected from the defect (the first part of equation 2), that is

$$h_{opt}(t)=\sum_{i=1}^{J}A_i s_0(t-t_i), \tag{4}$$

the SNR of the filter output $y(t)$ will be maximized. However, the defect echo $s_d(t)=\sum_{i=1}^{J}A_i s_0(t-t_i)$ is unknown, because the number of point reflectors J, the delays t_i and the amplitude factors A_i are all unknown, and may differ between samples. The goal of the optimal matched filter design is to tweak the parameters of the filter response $h(t)$ to approximate the unknown defect echo, so that the output SNR is maximized for a set of given training signals.

In mathematics, the above problem of optimal matched filter design can be formulated by the following constrained optimisation equations

$$\max_{A_i, t_i} \mathrm{SNR}\left[y(t)\right] \quad \text{subject to} \quad \begin{array}{l} y(t) = h(t) * x(t) \\ h(t) = \sum_{i=1}^{K} A_i s_0(t - t_i) \\ x(t) = \sum_{i=1}^{J} A_i s_0(t - t_i) + n(t) \end{array} \tag{5}$$

where *K* is the number of point reflectors considered in the approximation to reconstruct the defect echo, which is unknown *a prior* and is a design parameter. The defect echoes could be reconstructed using the actual excitation signals, which are accurate to account for the practical experimental uncertainties [12]; however the process could be troublesome requiring extra efforts and time. In this chapter, we present a more flexible and robust design method which uses simulated excitation signals taking into account of the operation parameters of the transducer and the transmitter-receiver system, to approximate the defect echoes. The experimental results from real industrial samples demonstrate that through the optimization and tuning procedure for the filter parameters, the same good results are obtained from this design routine.

From the problem formulation of optimal matched filter design (5), it demonstrates that the filter response determination consists of two steps: 1) To reconstruct the return echo by using the simulated excitation signal $s_0(t)$ and the design parameters K, A_i and t_i, $i = 1, 2, \ldots, K$; and 2) To tweak the above control parameters until the output SNR is maximized.

In the first step, with a particular set of control parameters K, A_i and t_i, $i = 1, 2, \ldots, K$, the return echoes are reconstructed, which are then employed in the matched filter (5) and the output SNR is calculated. The SNR will be increased or decreased depending on the parameters chosen in the filter. The second step of matched filter design involves tweak the above-mentioned parameters by searching the best values in the problem space, in order to achieve the maximum SNR gain.

Apparently, the search process is not trivial, and the dependence of the cost function $\mathrm{SNR}\left[y(t)\right]$ in equation (5) on the control variables A_i and t_i is supposed to be non-linear, high-dimensional and complicated. A numerical optimization paradigm such as the Genetic Algorithms (GA), Simulated Annealing (SA), or Particle Swarm Optimization (PSO) is in a good position to tackle this problem.

In this chapter, PSO is employed due to its robust and powerful optimisation capability, and simple implementation, as demonstrated in the literature [26]. PSO starts by initializing a population of particles in the search space. The position vector of the ith particle takes the form $p_i = \{A_{i1}, t_{i1}, A_{i2}, t_{i2}, \dots, A_{iK}, t_{iK}\}$. The fitness of particle p_i is computed using the flowchart in Figure 1. For more details about the update rules and equations of PSO paradigm and evolutionary algorithms, the readers may refer to [26-28] and the references therein.

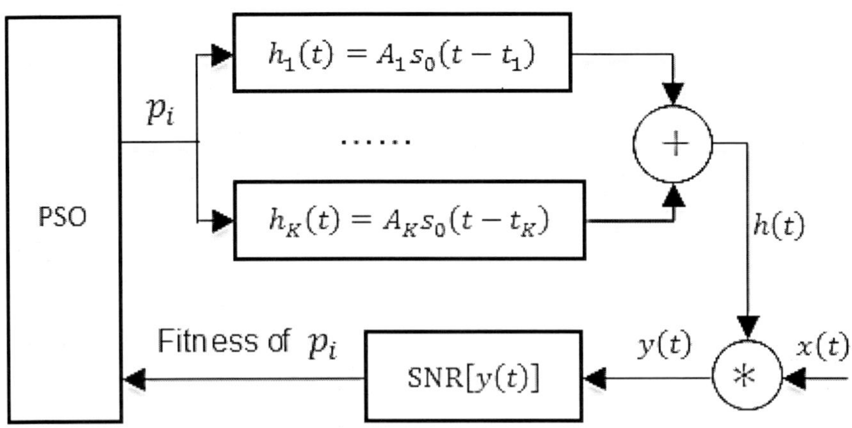

Figure 1. Determination of particle fitness in PSO paradigm.

3. Spatial Filtering and Adaptive Beamforming

3.1. Data Model and Problem Formulation

To properly formulate the problem and present the adaptive beamformer, a data model with point reflectors is considered in this section. Assume that there are J + 1 point scatters in the insonified region, each reflecting a signal, $s_j(t)$, j = 0, ..., J. A transducer array of N elements is considered. The time series received at the nth element is

$$x_n(t)=\sum_{j=0}^{J} g_{n,j}s_j\left(t-\tau_{n,j}\right)+v_n(t),\ n=1,\dots,N, \tag{6}$$

where $g_{n,j}$ represents the amplitude adjustment corresponding to the distance from reflector j to transducer element n and the gain of that element, $\tau_{n,j}$ is the time delay from reflector j to element n, and $v_n(t)$ is the noise on channel n.

Assume that the transducer array is focused on scatter 0 in reception through delaying each channel by $\tau_{n,0}$, n = 1, ..., N. Under this assumption, $s_0(t)$ is the desired signal and the other reflectors are sources of interference. The time-delayed signal received at channel n can be described as

$$x_n(t)=g_{n,0}s_0(t)+\sum_{j=1}^{J} g_{n,j}s_j\left(t+\tau_{n,0}-\tau_{n,j}\right)+v_n(t),$$
$$n=1,\dots,N, \tag{7}$$

The time-delayed array output vector $\mathbf{x}(t)$ is given by

$$\mathbf{x}(t)=[x_1(t)\ x_2(t)\ \dots\dots\ x_N(t)]^T, \tag{8}$$

where $[\bullet]^T$ denotes the transpose operator. The problem to be addressed in this section is to extract the desired signal $s_0(t)$ from the received echoes contaminated by the interference and additive noise using spatial filtering and adaptive beamforming.

3.2. Minimum Variance Adaptive Beamformer

Adaptive spatial filtering and beamforming is achieved by weighting the received echoes with data-dependent apodization, after each channel is appropriately delayed to focus at a point within the insonified media. The flow chart is illustrated in Figure 2. The output of the beamformer can be described as

$$z(t) = \mathbf{w}(t)^H \mathbf{x}(t), \tag{9}$$

where

$$\mathbf{w}(t) = [w_1(t)\ w_2(t)\ \dots\ \dots\ w_N(t)]^T \tag{10}$$

is a weight vector, and $[\bullet]^H$ stands for the operator of conjugate transpose. The minimum variance beamformer (also known as the Capon beamformer) seeks to minimize the power of the beamformed output $z(t)$ while maintaining unit gain on the focal point [29]. This constrained optimization problem can be formulated as

$$\min_{\mathbf{w}(t)} \mathbf{w}(t)^H \mathbf{R}(t)\mathbf{w}(t), \text{ subject to } \mathbf{w}(t)^H \mathbf{a} = 1, \tag{11}$$

where

$$\mathbf{R}(t) = E\left[\mathbf{x}(t)\mathbf{x}(t)^H\right] \tag{12}$$

is the data spatial covariance matrix, and

$$\mathbf{a} = [1\,1\ \ldots\ \ldots\ 1]^T \tag{13}$$

is an $N \times 1$ column vector of ones, because the data have already been delayed to focus on the point of interest. The optimal weight vector is the solution to (11), which is given by

$$\mathbf{w}_{opt}(t) = \frac{\mathbf{R}(t)^{-1}\mathbf{a}}{\mathbf{a}^H\mathbf{R}(t)^{-1}\mathbf{a}} \tag{14}$$

In practice, the ideal covariance matrix $\mathbf{R}(t)$ (12), which is computed from infinite number of data samples, used in (14) is not available, and has to be replaced by the sample covariance matrix. If the full matrix capture data acquisition [2] is used, the estimate can be obtained by

$$\hat{\mathbf{R}}(t) = \frac{1}{N}\sum_{k=1}^{N}\mathbf{x}^k(t)\mathbf{x}^k(t)^H, \tag{15}$$

where $\boldsymbol{x}^k(t)$ is the delayed array output vector when the kth array element is excited in transmission and the array is focused on the point of interest in reception.

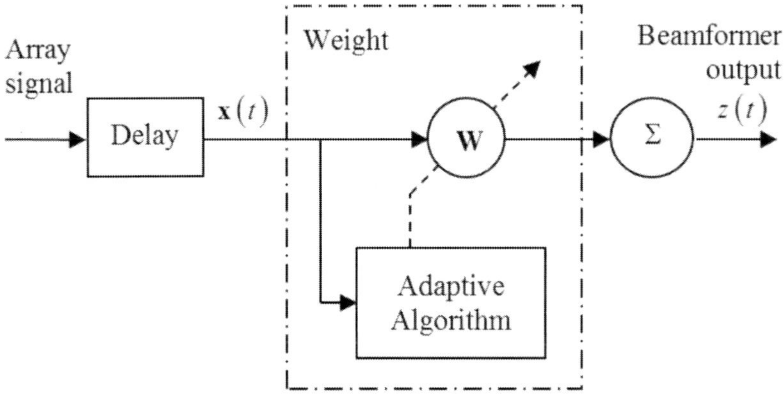

Figure 2. Flowchart of adaptive beamforming.

The minimum variance beamformer is an optimal spatial filter that maximizes the array output signal-to-interference-plus-noise ratio (SINR) [29]. However, its performance degrades in the presence of mismatches or errors like wrong assumptions of acoustic velocity or phase aberrations. This phenomenon is referred to as signal self-nulling, since the mismatches lead to targets appearing slightly out of focus and the beamformer will treat them as interference and try to minimize them [30, 31]. There exist several methods for increasing the robustness at the expense of resolution. Diagonal loading [32] is found to be a straightforward and effective approach in this application, where a constant η is added to the diagonal of the estimated covariance matrix

$$\hat{\mathbf{R}}_{DL}(t) = \hat{\mathbf{R}}(t) + \eta\mathbf{I}, \tag{16}$$

where $\mathbf{I}$ is a $N \times N$ identity matrix, and

$$\eta = \frac{1}{N} tr\left[\hat{\mathbf{R}}(t)\right], \tag{17}$$

where $tr[\bullet]$ denotes the trace operator. The amount of diagonal loading in (17) is proportional to the power in the received signals. However, η can be tweaked to make a trade-off between robustness and resolution, and a greater η leads to a more robust beamformer in (14), but the resolution is decreased.

4. Simulation and Experiment Results

In this section, we present the simulation and experiment results to evaluate the performances of the adaptive beamformer and the matched filter for interference and noise reduction, against the gold-standard Total Focusing Method (TFM) [2].

4.1. Simulation Studies

The schematic of a simulated specimen is shown in Figure 3, where multiple single and paired point reflectors are located at different depths and lateral positions. The material is assumed to be ideal, homogeneous and lossless with the wave propagation speed $c = 6{,}300\ \text{m}\cdot\text{s}^{-1}$. A 5 MHz linear transducer array is utilized in contact with the specimen upper surface, which consists of 64 omni-directional elements with the inter-element spacing 0.63 mm that is exactly the half wavelength in the medium at the centre frequency of the transducer. The output signal of each element is a five cycle, Gaussian windowed tone burst with a centre frequency of 5 MHz and a −6 dB bandwidth of 50%. The reflectors have a normalized unity amplitude with white Gaussian noise being added to the signals, corresponding to a SNR = 20 dB for a single echo and channel. The sampling rate is 100 MHz.

In the simulation studies, the capability of adaptive beamforming to image point reflectors and resolve closely spaced emitters is assessed and compared against that of the delay-and-sum beamformer. Figure 4 shows the images with a 70 dB dynamic range obtained by using TFM [Figure 4(a)] and adaptive beamforming in reception [Figure 4(b)] respectively, and the lateral profiles of the images at the depth of 30 mm are depicted in Figure 5.

It can be seen in Figure 4(a), using TFM, the sidelobe indications are pretty visible, especially in the regions close to the main lobe, which potentially hinder the detection of surrounding smaller or weaker reflectors and limit the capability to resolve clustered reflectors. When adaptive beamforming is applied, the image quality is significantly enhanced in terms of clutter noise, image resolution, contrast, and dynamic range as shown in Figure 4(b). The lateral profiles in Figure 5 indicate that the main lobe width is more than 50% smaller (−6 dB main lobe width of TFM being 0.93 mm and that of adaptive beamforming being 0.42 mm) and the noise is reduced by more than 30 dB, when adaptive beamforming is applied.

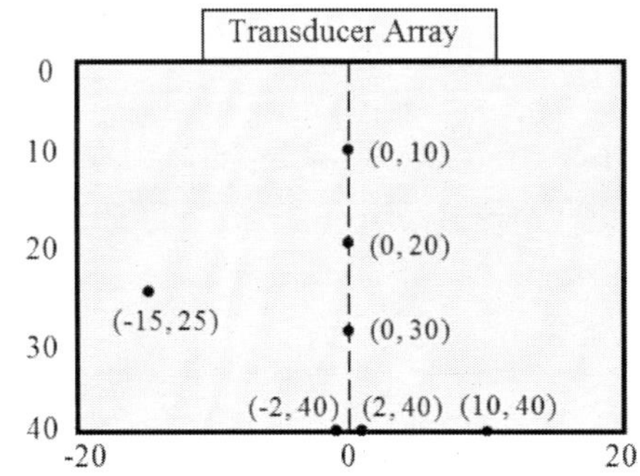

Figure 3. Schematic of simulated sample.

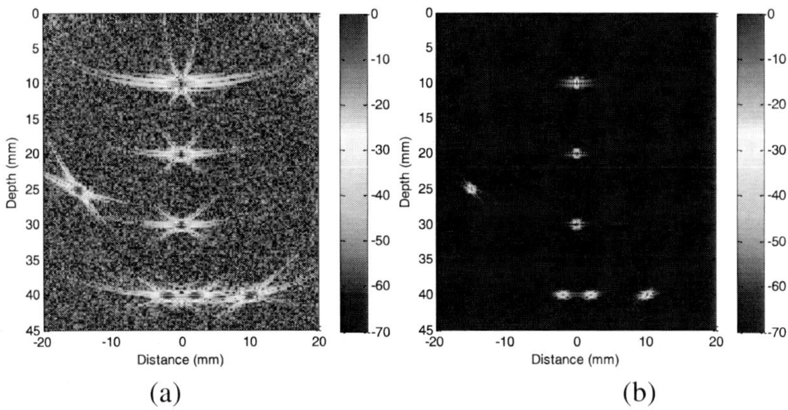

Figure 4. Images of simulated sample. (a) TFM, and (b) Adaptive beamforming.

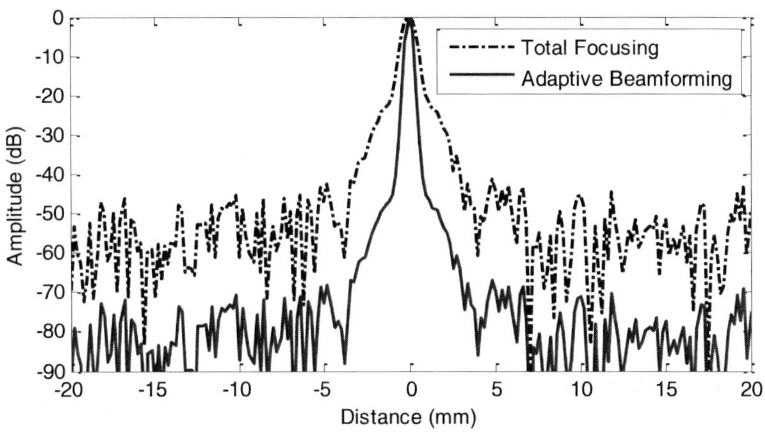

Figure 5. Lateral profiles of images at depth of 30 mm.

4.2. Experimental Verification

The experimental apparatus consists of three components, namely, test sample, array transducer, phased array control system with a personal computer for excitation and signal processing. A 128-element transducer array with 0.7 mm element pitch and 5 MHz central frequency (Vermon, Tours, France) is utilized in contact with the test sample upper surface with gel coupling, as shown in Figure 6(a). An OPEN ultrasonic phased array control system with 128 independent parallel channels and 16-bit resolution (LeCouer, Chuelles, France) is connected with the transducer array for excitation and data acquisition, as shown in Figure 6(b). A personal computer is connected to the OPEN system to control the excitation sequence and record the return signals for post processing and imaging. A MATLAB (The MathWorks, Natick, MA) routine is developed to implement the Full Matrix Capture (FMC) data acquisition, where each transducer element is excited sequentially and the echoes received by all the array elements are recorded [2]. A complete FMC data set is composed of N^2 A-scan waveforms, where N is the number of array elements. FMC data are recorded at a sampling rate of 100 MHz, and each A-scan waveform is band-pass filtered to remove the DC drift and high frequency noise.

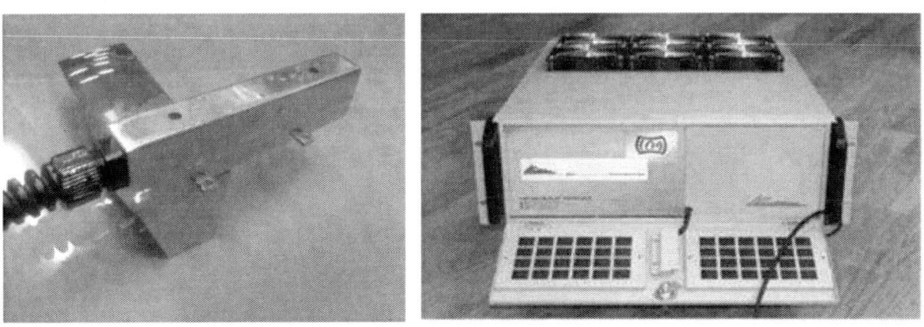

(a) (b)

Figure 6. Experiment apparatus. (a) Transducer array and test sample, and (b) OPEN phased array control system.

Table 1. PSO parameters

PSO Parameters	Value
K	15
Number of Particles	150
Number of Iterations	500
c_1	0.4
c_2	1.2
Inertia Weight	0.8

In the experiments, the matched filter response is formulated using equation (5) with K reflectors and all the unknown parameters are captured in a vector with a dimension of $2K \times 1$, consisting of A_i and t_i, $i=1, 2, \ldots, K$. K is pre-determined and set $K=15$, which seems to be not highly sensitive to the performance, provided that a reasonably large K is employed. The PSO paradigm is employed to tackle the problem at hand, because PSO is simple to implement, but it is as powerful as other popular evolutionary algorithms like the Genetic Algorithms, and is robust to the selection of control parameters [28]. The key parameters used in the PSO routine are summarized in Table 1, and the convergence is observed in all the experiments within the maximum number of iterations.

A test sample with a thickness of around 150 mm made of INCONEL Alloy 617 is employed in the experiments. As shown in the datasheet, INCONEL Alloy 617 is a solid-solution, nickel-chromium-cobalt-molybdenum alloy with an exceptional combination of high-temperature strength and oxidation resistance [33]. Alloy 617 is widely used in a variety of key industrial sectors like Energy, Oil and Gas, and Nuclear for components such as ducting and combustion cans. Due to the size of microstructure grain, Alloy 617 is demonstrated to be highly scattering to 5 MHz ultrasound, and results in significant grain noise and pretty low SNR.

Figure 7(a) shows the A-scan waveform of the echo at Element 40 when the same transducer element is excited, the reflection from a major defect happens around the samples of 1630-1730. Due to the dominant grain noise, it is almost impossible to identify this reflection, and the SNR is measured to be -9 dB. Figure 8(a) demonstrates the image obtained with TFM using the original FMC dataset in a dynamic range of 40 dB. The image is pretty noisy and the back-wall is only partially visible at the depth of 147 mm.

Figure 7(b) illustrates the same echo at Element 40 as in Figure 7(a), but processed with the optimized matched filter. Comparing to the original signal in Figure 7(a), it is evident that the defect echo has been enhanced and the grain noise has been suppressed with matched filtering, and it becomes much more confident to detect the defect from thresholding the amplitude. In terms of quantitative comparison, the SNR is increased to 7 dB, and the matched filter observes 16 dB SNR gain for the A-scan signal. To form the image in Figure 8(b), firstly, each individual A-scan waveform in the FMC data set is matched filtered, and then the processed FMC data set is imaged with TFM. Comparing to the image in Figure 8(a), it is evident that the clutter noise is greatly reduced, while the back wall reflection is further enhanced.

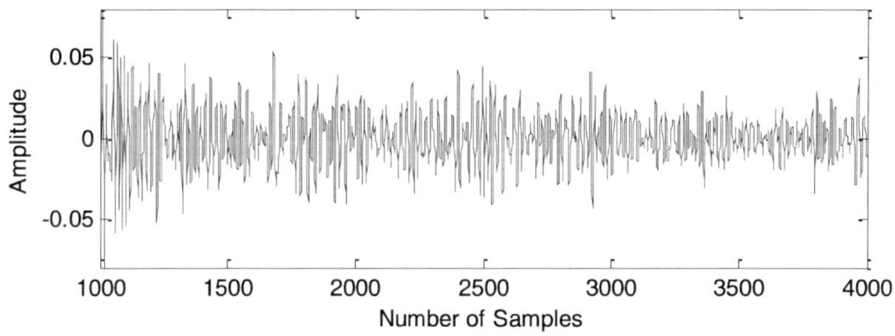

(a)

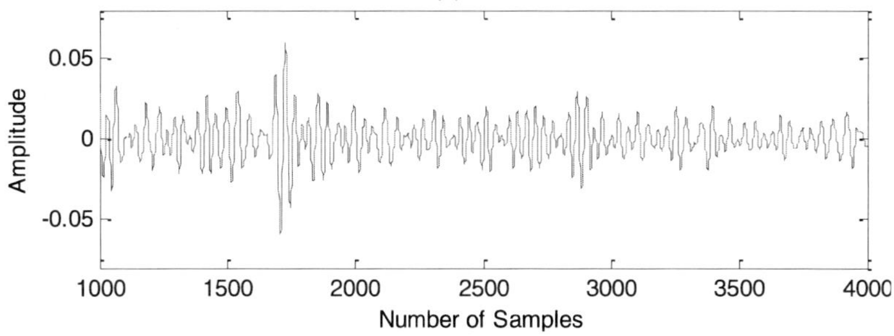

(b)

Figure 7. A-scan waveforms at Element 40. (a) Original, and (b) Matched filtered.

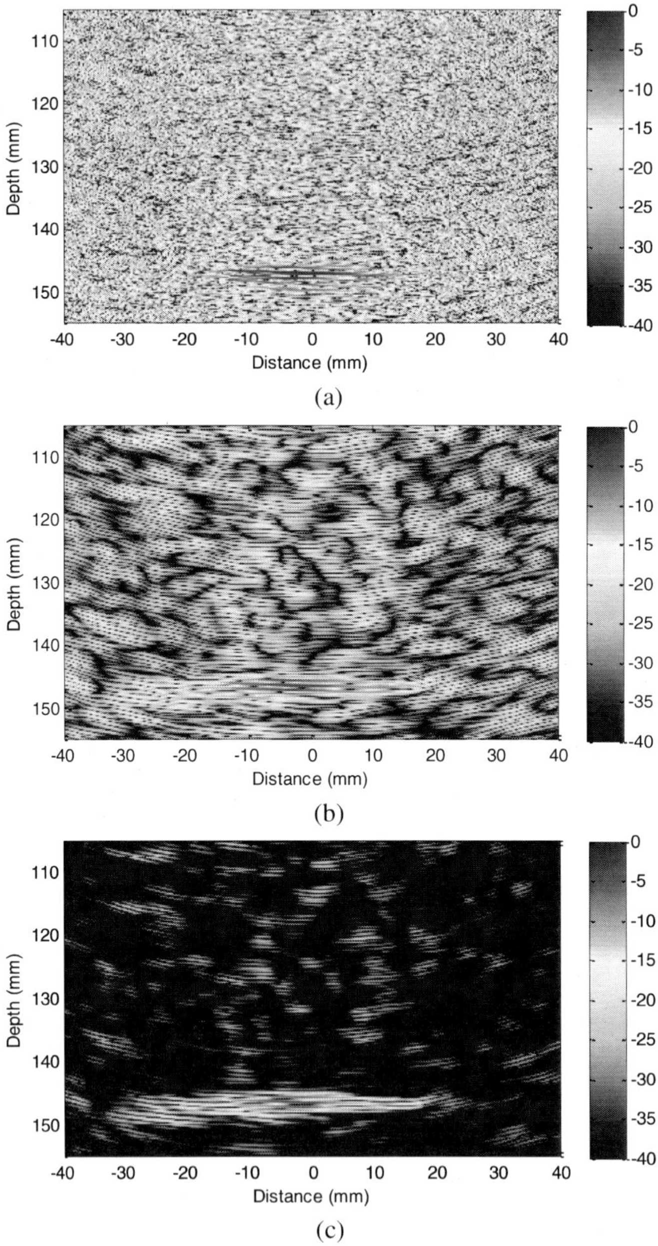

Figure 8. Images of Alloy 617 test block. (a) TFM with original FMC data, (b) TFM with matched filtered FMC data, and (c) Adaptive beamforming with matched filtered FMC data.

The matched filter works in the temporal-spectral domain, and the adaptive beamformer exploits the spatial diversity along the transducer array aperture. Apparently, there is a potential to combine these two techniques since they work in different domains and may complement with each other for noise reduction. Figure 8(c) shows the image obtained with the minimum variance beamformer using the matched filtered FMC data set. It is evident that the clutter noise is further reduced, and the back-wall reflection is greatly enhanced in term of the strength as well as the visible lateral length, due to the exploitation of both temporal and spatial characteristics.

CONCLUSION

This chapter addresses the challenging problem of noise reduction in ultrasonic NDE, and focuses on matched filtering and adaptive beamforming for enhanced SNR and robust defect detection. Matched filtering is demonstrated to be a powerful and effective tool for the problem at hand, and a flexible approach to optimal matched filter design is presented, where the simulated excitation signals are utilised with the control parameters being tweaked and optimized taking into account the actual responses of transducers, the transceiver channels, and the test sample characteristics. The filter is tuned to match the unknown defect echoes instead of the excitation signal, and the response is dependent on the material properties. Minimum variance beamforming demonstrates significant performance improvement over the standard delay-and-sum operation, and diagonal loading further increases its robustness in ultrasonic NDE. With adaptive beamforming applied spatially to the observations across the probe aperture, the spurious echoes and clutter noise are significantly reduced. Simulation studies confirm that the noise level is reduced by more than 30dB and the lateral resolution is improved by more than 50% when adaptive beam-forming is employed in the scenario of multiple point reflectors. Experiments on an industrial sample of INCONEL Alloy 617 are conducted and the results verify the advantages and robustness of the approaches when applied separately, and confirm the greater benefits when they are combined to complement with each other as they work in temporal- and spatial-domains respectively. It can be concluded that, the approaches offer a great potential to inspect coarse-grained materials, enhance sensitivity to smaller flaws, increase imaging resolution and contrast, and improve the definition of defect size and shape in NDE.

REFERENCES

[1] Thompson, RB; Thompson, DO. "Ultrasonics in nondestructive evaluation," *Proc. of the IEEE*, vol. 73, pp. 1716-1755, Dec. 1985.

[2] Li, M; Hayward, G. "Ultrasound non-destructive evaluation (NDE) imaging with transducer arrays and adaptive processing," *Sensors*, vol. 12, pp. 42-54, Jan. 2012.

[3] Papadakis, EP. "Ultrasonic attenuation caused by scattering in polycrystalline metals," *J. Acooust. Soc. Amer*, vol. 37, pp. 703-710, 1965.

[4] Neal, SP; Speckman, PL; Enright, MA. "Flaw signature estimation in ultrasonic nondestructive evaluation using the Wiener filter with limited prior information," *IEEE Trans. Ultrason., Ferroelect., Freq. Contr.*, vol. 40, pp. 347-353, July 1993.

[5] Izquierdo, MAG; Hernández, MG; Graullera, O; Ullate, LG. "Time-frequency Wiener filtering for structural noise reduction," *Ultrasonics*, vol. 40, pp. 259-261, May 2002.

[6] Amir, I; Bilgutay, NM; Newhouse, VL. "Analysis and comparison of some frequency compounding algorithms for the reduction of ultrasonic clutter," *IEEE Trans. Ultrason. Ferroelectr. Freq. Contr.*, vol. 33, pp. 402-411, 1986.

[7] Qi, T; Bilgutay, NM. "Statistical analysis of split spectrum processing for multiple target detection," *IEEE Trans. Ultrason., Ferroelect., Freq. Contr.*, vol. 45, pp. 251-256, Jan. 1998.

[8] Ho, K; Li, M; O'Leary, R; Gachagan, A. "Application of frequency compounding to ultrasonic signals for the NDE of concrete," in *Review of Progress in Quantitative Nondestructive Evaluation*, eds. D. O.

Thompson and D. E. Chimenti, (American Institute of Physics 1430, Melville, NY), vol. 31, pp. 1508-1515, 2012.

[9] Gongzhang, R; Li, M; Lardner, T; Gachagan, A. "Robust defect detection in ultrasonic nondestructive evaluation (NDE) of difficult materials," *Proc. 2012 IEEE International Ultrasonics Symposium (IUS)*, pp. 467-470, Oct. 2012.

[10] Gongzhang, R; Li, M; Xiao, B; Lardner, T; Gachagan, A. "Robust frequency diversity based algorithm for clutter noise reduction of ultrasonic signals using multiple sub-spectrum phase coherence," in *Review of Progress in Quantitative Nondestructive Evaluation*, eds. D. E. Chimenti, L, J. Bond and D. O. Thompson, (American Institute of Physics 1581, Melville, NY), vol. 33, pp. 1948-1955, 2014.

[11] Xiao, B; Li, M; Gongzhang, R; O'Leary, R; Gachagan, A. "Image de-noising via spectral distribution similarity analysis for ultrasonic non-destructive evaluation," in *Review of Progress in Quantitative Nondestructive Evaluation*, eds. D. E. Chimenti, L, J. Bond and D. O. Thompson, (American Institute of Physics 1581, Melville, NY), vol. 33, pp. 1941-1947, 2014.

[12] Li, M; Hayward, G. "A rapid approach to speckle noise reduction in ultrasonic non-destructive evaluation using matched filters," *Proc. 2014 IEEE International Ultrasonics Symposium (IUS)*, pp. 45-48, Sept. 2014.

[13] Li, M; Hayward, G. "Optimal matched filter design for ultrasonic NDE of coarse grain materials," in *Review of Progress in Quantitative Nondestructive Evaluation*, eds. D. E. Chimenti and L. J. Bond, (American Institute of Physics 1430, Melville, NY), vol. 35, pp. 020011 1-9, 2016.

[14] Li, M; Hayward, G; He, B. "Adaptive array processing for ultrasonic non-destructive evaluation," *Proc. 2011 IEEE International Ultrasonics Symposium (IUS)*, pp. 2029-2032, Oct. 2011.

[15] Li, PC; Li, ML. "Adaptive imaging using the generalized coherence factor," *IEEE Trans. Ultrason., Ferroelect., Freq. Contr.*, vol. 50, pp. 128-141, Feb. 2003.

[16] Lardner, T; Li, M; Gongzhang, R; Gachagan, A. "A new speckle noise suppression technique using cross-correlation of array subapertures in ultrasonic NDE of coarse grain materials," in *Review of Progress in Quantitative Nondestructive Evaluation*, eds. D. O. Thompson and D. E. Chimenti, (American Institute of Physics 1581, Melville, NY), vol. 32, pp. 865-871, 2013.

[17] Lardner, T; Li, M; Gachagan, A. "Using phase information to enhance speckle noise reduction in the ultrasonic NDE of coarse grain materials," in *Review of Progress in Quantitative Nondestructive Evaluation*, eds. D. E. Chimenti, L, J. Bond and D. O. Thompson, (American Institute of Physics 1581, Melville, NY), vol. 33, pp. 1061-1069, 2014.

[18] Gjessing, D. "Adaptive techniques for radar detection and identification of objects in an ocean environment," *IEEE Journal of Oceanic Engineering*, vol. 6, pp. 5-17, Jan. 1981.

[19] Srinivasan, K; Chiou, CP; Thompson, RB. "Ultrasonic flaw detection using signal matching techniques," in *Review of Progress in Quantitative Nondestructive Evaluation*, eds. D. O. Thompson and D. E. Chimenti, (Plenum Press, NY), vol. 14, pp. 711-718, 1995.

[20] Ruiz-Reyes, N; Vera-Candeas, P; Curpian-Alonso, J; Mata-Campos, R; Cuevas-Martinez, JC. "New matching pursuit-based algorithm for SNR improvement in ultrasonic NDT," *NDT&E International*, vol. 38, pp. 453-458, June 2005.

[21] Li, M; Lu, Y. "Improving the performance of GA-ML DOA estimator with a resampling scheme," *Signal Processing*, vol. 84, pp. 1813-1822, Oct. 2004.

[22] Li, M; Lu, Y. "Source bearing and steering-vector estimation using partially calibrated arrays," *IEEE Trans. Aerosp. Electron. Syst.*, vol. 45, pp. 1361-1372, Oct. 2009.

[23] Li, M; Lu, Y. "Optimal direction finding in unknown noise environments using antenna arrays in wireless sensor networks," *Proc. 7th International Conference on Intelligent Transportation Systems Telecommunications (ITST2007)*, pp. 332-337, June 2007.

[24] Li, M; Ho, KS; Hayward, G. "Accurate angle-of-arrival measurement using particle swarm optimisation," *Wireless Sensor Network*, vol. 2, pp. 358-364, May 2010.

[25] Holmes, C; Drinkwater, B; Wilcox, P. "Post-processing of the full matrix of ultrasonic transmit-receive array data for non-destructive evaluation," *NDT&E International*, vol. 38, pp. 701-711, Aug. 2005.

[26] Eberhart, R; Kennedy, J. "A new optimizer using particle swarm theory," *Proc. 6th International Symposium on Micro Machine and Human Science*, pp. 39-43, Oct. 1995.

[27] Li, M; Lu, Y. "A refined genetic algorithm for accurate and reliable DOA estimation with a sensor array," *Wireless Personal Communications*, vol. 43, pp. 533-547, Oct. 2007.

[28] Li, M; Lu, Y. "Maximum likelihood DOA estimation in unknown colored noise fields," *IEEE Trans. Aerosp. Electron. Syst.*, vol. 44, pp. 1079-1090, Jul. 2008.

[29] Capon, J. "High-resolution frequency-wavenumber spectrum analysis," *Proc. of the IEEE*, vol. 57, pp. 1408-1418, 1969.

[30] Synnevag, JF; Austeng, A; Holm, S. "Adaptive beamforming applied to medical ultrasound imaging," *IEEE Trans. Ultrason. Ferroelectr. Freq. Contr.*, vol. *54*, pp. 1606-1613, Aug. 2007.

[31] Li, M; Lu, Y. "Dimension reduction for array processing with robust interference cancellation," *IEEE Trans. Aerosp. Electron. Syst.*, vol. *42*, pp. 103-112, Jan. 2006.

[32] Li, J; Stoica, P; Wang, Z. "On robust Capon beamforming and diagonal loading," *IEEE Trans. Sign. Process.*, vol. *51*, pp. 1702-1715, 2003.

[33] INCONEL 617 *Technical Data*, http://www.hightempmetals.com/techdata/hitempInconel617data.php, accessed in February 2018.

In: Noise Reduction
Editor: Marc Farrow
ISBN: 978-1-53613-541-1

Chapter 5

THE EVALUATION OF AN ACTION PLAN FOR REDUCTION OF ENVIRONMENTAL NOISE IN RIGA AGGLOMERATION

Janis I. Dundurs, PhD[1,*] and Miervaldis Lacis, PhD[2]
[1]Department of Occupational and Environmental Medicine,
Riga Stradins University, Riga, Latvia
[2]Department of Housing and Environment,
Riga City Municipality, Riga, Latvia

ABSTRACT

The purpose of this article is to show how Riga City Municipality has managed to fulfil the main goal of the 'Directive 2002/49/EC of the European Parliament and of the Council of 25 June 2002 relating to the assessment and management of environmental noise' – to reduce the number of environmental noise affected people in the Riga agglomeration. According to Directive 2002/49/EC, for this purpose the Riga City Municipality drew up the 'Action Plan for Reduction of Environmental

[*] Corresponding Author Email: janis.dundurs@rsu.lv.

Noise in Riga Agglomeration' in 2014, which has been approved by the Riga City Council on July 8, 2014. This Action Plan scheduled the measures to reduce or to prevent the harmful effects of environmental noise emitted by road traffic, especially on major roads, by rail traffic and airplanes, as well as by industrial sites and open air industrial equipment.

As the Directive 2002/49/EC obligates the review and to revise Action Plans at intervals of at least every five years, Riga City Municipality timely carried out the efforts to assess the data about noise affected people as well as gains and possible losses after the 'Action Plan for Reduction of Environmental Noise in Riga Agglomeration' (2014) was implemented.

The Action Plan contains the medium-term and long-term activities devoted to the reduction of noise pollution in the dwelling areas, in public parks and other quiet areas in Riga City. All planned measures belong to noise reduction technical solutions – sound barriers (hedgerows and tree rows included), city transport infrastructure pavement and tram railway quality improvement, cargo train speed limitation on inner city railways and building façade sound insulation ability improvement by glass panels as well. It was estimated to spend some 90,000 EUR for medium-term measures but for long-term measures - more than 10 million EUR.

Keywords: environmental noise, action plan, agglomeration, noise barrier

AIMS AND SCOPE OF THE EVALUATION

The document created by the Riga City Municipality, "Evaluation of Action Plan for Reduction of Environmental Noise in Riga Agglomeration" (year 2014) on fulfilling planned tasks and achieving main goals or to carry over the tasks to the new "Action Plan for Reduction of Environmental Noise in Riga Agglomeration (on period from 2017 to 2022)" clearly shows how effective or non-effective are the previously chosen noise reduction methods and applications. The first Action Plan for Reduction of Environmental Noise in Riga Agglomeration (year 2014) as the very first document of that kind in municipality, should now be considered as a useful attempt to find the ways and benefits of how to manage environmental noise in Riga agglomeration. The gains and losses of each chosen noise reduction method, on the other hand, help us to decide whether this method deserves to be carried over and enclosed in an Action Plan for the future.

Description of Riga Agglomeration and Noise Sources

The Riga agglomeration consists of Riga city with total area of 304.5 km^2. The number of inhabitants in the agglomeration – 650,500, average density – 2,136 inhabitants/km^2. Strategic noise maps encompass 4 types of sources – road transport (5,850 street sections with total length 1,180 km), rail transport (395 km of railroad and tram lines), industrial sources (268 objects with total area 11.6 km^2) and aircraft (67,200 flights to and from Riga International Airport).

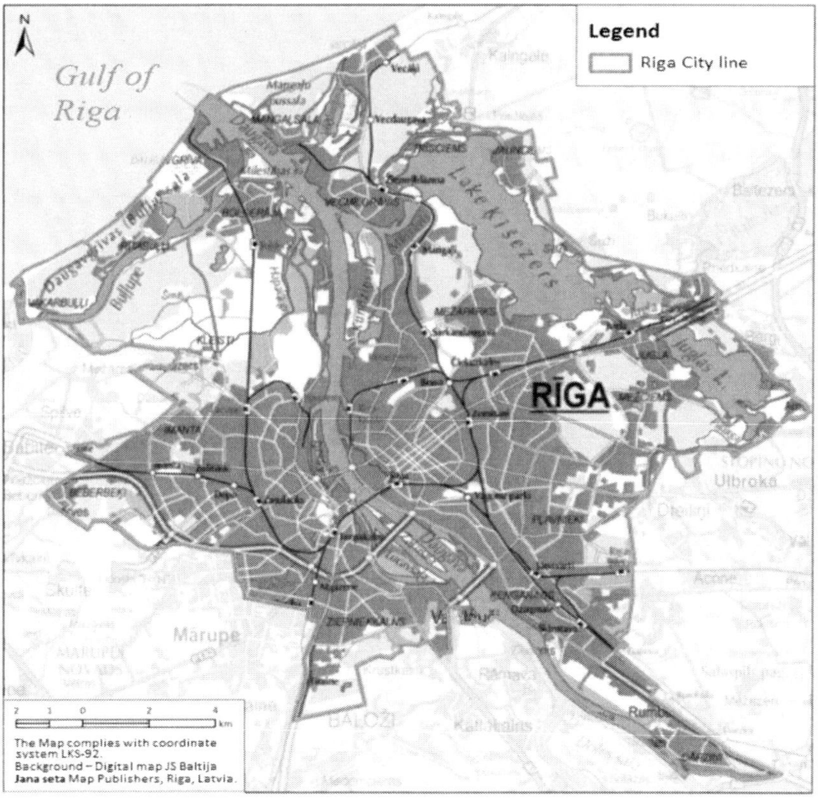

Figure 1. Riga Agglomeration.

Noise assessment and management on the EU level is carried out in accordance with the "Directive 2002/49/EC of the European Parliament and of the Council of 25 June 2002 relating to the assessment and management of environmental noise" [1]. The directive prescribes that the EU member states must develop strategic noise maps and noise reduction action plans for all agglomerations with more than 100,000 inhabitants. In accordance with the requirements of the directive the review of strategic noise maps and action plans must be carried out every five years. These requirements of the directive are transposed in the Latvian law "On Pollution", delegating the development of detailed environmental noise assessment and management procedures to the Cabinet of Ministers. The Cabinet of Ministers has adopted Regulations No. 16 "Noise assessment and management procedures," which sets noise assessment and management requirements and the noise limit values.

Based on 2015 strategic noise map data, it has been concluded that the main source of noise in Riga agglomeration is road transport. Approximately 496,000 or 76% of inhabitants are exposed to road transport noise levels above 55 dB (A) during day time. Approximately 560,000 inhabitants are exposed to road transport noise above 50 dB (A) during evening time. Approximately 538,000 inhabitants are exposed to road transport noise above 45 dB (A) at night.

Approximately 14,000 or 2% of inhabitants are exposed to rail transport-caused noise levels above 55 dB (A) at day time. Rail transport noise above 5 dB (A) is affecting approximately 40,000 inhabitants during evening time. Approximately 61,000 of inhabitants are exposed to rail transport noise levels above 45 dB (A) at night. Approximately 367 inhabitants are exposed to air traffic noise levels above 55 dB (A) at day time. Approximately 694 inhabitants are exposed to air traffic noise above 50 dB (A) during evening time. Approximately 895 inhabitants are exposed to air traffic noise levels above 45 dB (A) at night. Day-time industrial activity-caused noise levels above 55 dB (A) affect approximately 2,500 inhabitants. Approximately 9,000 inhabitants are exposed to industrial noise levels above 50 dB (A) during evening time. Approximately 21,000 inhabitants are exposed to industrial noise above 45 dB (A) at night. Analyzing noise level compliance with environmental noise limit values, it was concluded that day-time noise limit values are exceeded in 39.9 km^2 or 13% of the total Riga area, evening time noise limit values are exceeded in 57.25 km^2 or 19% of the total area, and night time limit values are exceeded in 52.4 km^2 or 17% of the total area.

THE MAIN PLANNED ACTIONS FOR REDUCING THE ENVIRONMENTAL NOISE IN RIGA AGGLOMERATION AND THE RESULTS OF IMPLEMENTATION

Approved on July 8, 2014 by the Riga City Council "Action Plan for Reduction of Environmental Noise in Riga Agglomeration" encompassed the main planned activities, such as: technical measures for the reduction of environmental noise emissions from noise sources and noise propagation, railway traffic regulation adjustment in Riga city lines and technical measures for noise emission reduction in noise sensitive buildings where humans were exposed. Developing events in the field of current road and bridge construction and the need for quick solutions to provide reasonable noise protective measures led to the most effective environmental noise reduction technique – noise barriers. Therefore, the choice fell on a more beneficial method of noise pollution reduction in the city – the noise barriers were erected in the sound propagation track between the noise source and noise receptor if there was enough space for them.

The tasks were grouped depending on executors. Part of tasks related to motorways and streets inside the city line were addressed to Riga Municipality departments and subordinate enterprises. The other part of planned tasks concerning railway links inside the city were addressed to the holding company SJSC "Latvijas dzelzceļš" (Latvian Railways).

Riga Municipality Traffic Department was responsible for the erection of sound barriers along the major city roads - inner highways with intensive traffic in two places: DIENVIDU TILTS (South Bridge) Highway in KATLAKALNS and Kārļa Ulmaņa gatve overpass across Riga – Jelgava railway.

Implemented Tasks for the Reduction of Environmental Noise in Riga Agglomeration

Noise Barriers, 3 M High on the Railway Overpass at Riga City Inner Highway Kārļa Ulmaņa Gatve in Torņakalns District

(Source: <maps.google.com>).

Figure 2. Noise barrier on railway overpass at Torņakalns district.

The length – 846 m on both sides of the two direction road. Noise level reduction in the surrounding area of railway overpass (calculated):

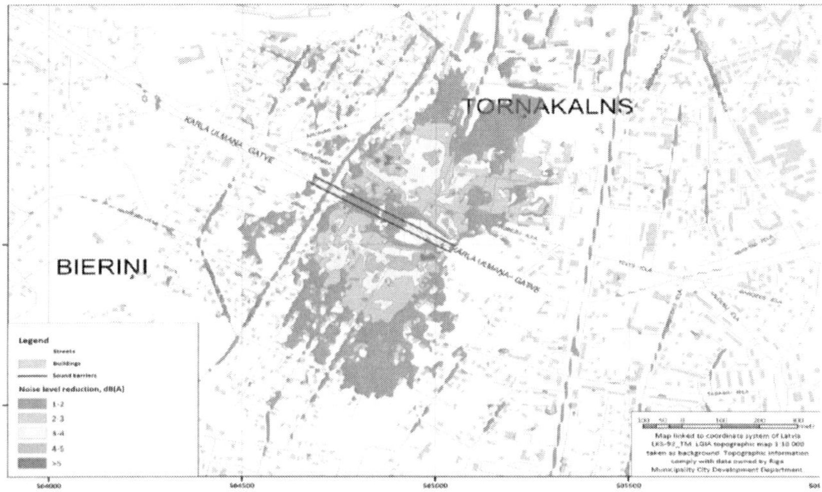

Figure 3. The influence of sound barriers on noise levels alongside highway Kārļa Ulmaņa gatve.

Table 1. Number of inhabitants subjected to certain noise levels before and after the noise barriers were erected [2], [3]

Sound barriers alongside highway Kārļa Ulmaņa gatve		Number of inhabitants at day time		Number of inhabitants in the evening		Number of inhabitants at night	
		before action	after the action took effect	before action	after the action took effect	before action	after the action took effect
Yearly average equivalent noise levels in dB(A)	<45	96	0	96	96	1,192	1,341
	45-50	282	268	539	419	3,187	3,354
	50-55	1,870	2 028	2 845	3 176	3,909	3,829
	55-60	3,982	4 343	4 037	4 324	2,973	2,974
	60-65	3,713	3 518	3 248	2 858	1,917	1,715
	65-70	2,360	2 283	2 248	2 248	115	80
	70-75	969	829	280	172	0	0
	>75	21	24	0	0	0	0
Total inhabitants:		13,293	13,293	13,293	13,293	13,293	1,3293
Inhabitants at L < 60dB:			409		498		237
Gain total at L < 60dB:		1,144					
Gain total percentage: 9%							

As it follows from the data in Table 1, 1,144 inhabitants in the surrounding territory have experienced a life quality improvement - the yearly average equivalent noise levels below 60 dB (A). Among them, 100 inhabitants got even more lucky, being exposed to noise levels below 50 dB (A) at night.

Noise Barriers 3 M and 4 M High on the Railway Overpass at Riga City Inner Highway in Krasta Masīvs District

(Source: <maps.google.com>).

Figure 4. Noise barrier on railway overpass at Krasta Masīvs district.

Total length approximately 2,200 m. Noise level reduction in the surrounding area of railway overpass (calculated):

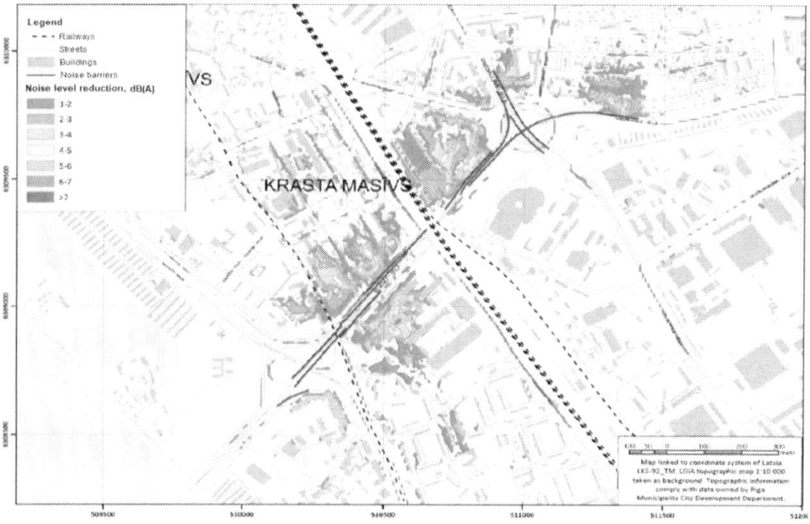

Figure 5. The influence of sound barrier on noise levels on railway overpass in krasta masīvs district.

Table 2. Number of inhabitants subjected to certain noise levels before and after the noise barriers were erected [2], [3]

Sound barriers along railway overpass at KRASTA MASĪVS		Number of inhabitants at day time		Number of inhabitants in the evening		Number of inhabitants at night	
		before action	after the action took effect	before action	after the action took effect	before action	after the action took effect
Yearly average equivalent noise levels in dB(A)	<45	215	0	356	141	4,940	4,952
	45-50	2,438	2,244	2,948	3,047	4,306	4,466
	50-55	3,883	3,798	4,367	4,613	7,207	7,691
	55-60	5,093	6,491	6,751	7,007	9,648	9,651
	60-65	6,897	8,860	10,386	11,068	2,508	1,849
	65-70	9,379	7,073	3,790	2,733	0	0
	70-75	704	143	11	0	0	0
Total inhabitants:		28,609	28,609	28,609	28,609	28,609	28,609
Inhabitants at L < 60dB:			494		130		656
Gain total at L < 60dB:		292					
Gain total percentage: 1%							

As it follows from the data in Table 2, 292 inhabitants in the surrounding territory have experienced a life quality improvement – the yearly average equivalent noise levels below 60 dB (A). Among them, 172 inhabitants got even more lucky being exposed to noise levels below 50 dB (A) at night.

THE TASK CARRIED OVER TO THE NEW "ACTION PLAN FOR REDUCTION OF ENVIRONMENTAL NOISE IN RIGA AGGLOMERATION (PERIOD FROM 2017 TO 2022)"

Dienvidu Tilts (South Bridge) Highway in Katlakalns district

The length is approximately 2,000 m on both sides of the two direction road. Noise level reduction in surrounding area (calculated).

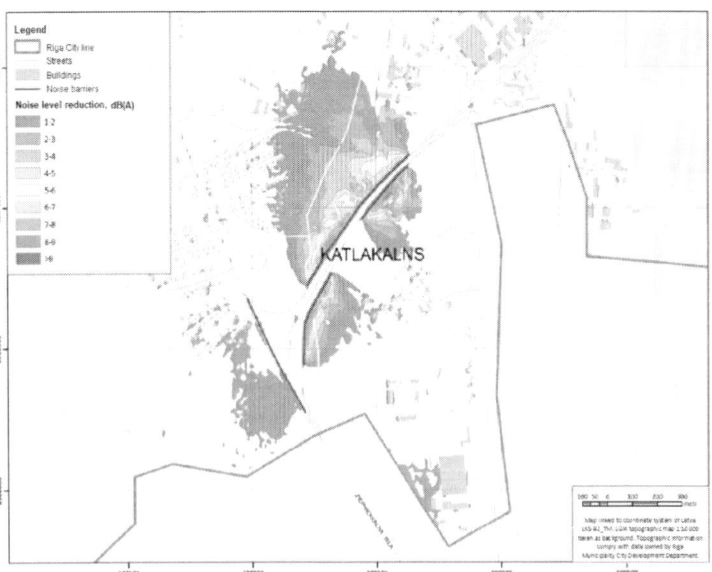

Figure 6. Sound barrier influence on noise levels alongside the Dienvidu Tilts highway.

Table 3. Number of inhabitants subjected to certain noise levels before and after the noise barriers were erected [2], [3]

Sound barriers alongside with DIENVIDU TILTS highway		Number of inhabitants at day time		Number of inhabitants in the evening		Number of inhabitants at night	
		before action	after the action took effect	before action	after the action took effect	before action	after the action took effect
Yearly average equivalent noise levels in dB(A)	<45	0	0	0	0	238	266
	45-50	1	9	15	171	792	864
	50-55	493	542	752	812	750	870
	55-60	814	1325	983	1001	198	81
	60-65	625	175	208	94	256	170
	65-70	211	106	208	121	17	0
	70-75	107	94	85	52	0	0
Total inhabitants:		2,251	2,251	2,251	2,251	2,251	2,251
Inhabitants at L < 60dB:			568		234		103
Gain total at L < 60dB:		905					
Gain total percentage: 40%							

As shown in Table 3,905 inhabitants in the surrounding territory can expect a life quality improvement – the yearly average equivalent noise levels below 60 dB (A). Among them, 100 inhabitants could become even more lucky, being exposed to noise levels below 50 dB (A) at night.

CONCLUSION

The paper "Evaluation of Action Plan for Reduction of Environmental Noise in Riga Agglomeration (year 2014) on fulfilling planned tasks and achieving main goals or to carry over the tasks to the new Action Plan for Reduction of Environmental Noise in Riga Agglomeration (on period from 2017 to 2022)" shows how effective or non-effective are the previously chosen noise reduction methods and applications. As mentioned above, the first Action Plan for Reduction of Environmental Noise in Riga Agglomeration (year 2014) comprises planned activities, addressed mainly to Riga Municipality departments and subordinate enterprises and to the holding company "Latvijas dzelzceļš" (Latvian Railways).

Unfortunately, the choice to oblige Latvian Railways to reduce train speed on inner cities railway links and planting of trees and bushes along inner cities railway lines were unsuccessful. The planned reduction of the train speed inside the city met disinterest from the Latvijas dzelzceļš (*Latvian Railways*). Latvian Railways considered all planned activities ineffective and rejected them completely. Riga Municipality Housing and Environment Department, also, after more scrupulous studies found out that the building façade sound isolation quality improvement by glass revetment gave little benefit in sound isolation quality (2 up to 4 dB) and required great expense. For instance, to 'coat' the medium size building's most exposed wall to a 'double skin' façade requires approximately 2 million euros. As a result, the noise barriers were widely used as the most effective and inexpensive measure for the reduction of noise emission from noise sources.

Since, in accordance with the Directive 2002/49/EC, the main goal of every planned activity is to reduce the number of noise affected people, the evaluation comprised an assessment of what number of Riga City inhabitants are subjected to environmental noise yearly average levels – the noise indicators – before and after the noise reduction tasks were implemented or planned to be undertaken. It is sad to recognize that not a very large amount of people can enjoy the results of the noise protective efforts. For instance, the noise barrier on the railway overpass at Torņakalns district improves the life quality to only 9% of district inhabitants. The noise barrier on the railway overpass at Riga City inner highway in Krasta Masīvs district is an improvement to only 1% of district inhabitants. Only the planned noise barrier at Dienvidu Tilts (South Bridge) Highway in Katlakalns district is expected to improve the life quality of up to 40% of district inhabitants.

REFERENCES

[1] The European Parliament and the Council of the European Union. DIRECTIVE 2002/49/EC of the European Parliament and of the Council of 25 June 2002 relating to the assessment and management of environmental noise. *Official Journal of the European Communities*, L 189/12.

[2] SIA Latvian, Estonian & Lithuanian Environment (2017). *"Ziņojums par Rīcības plāna (2014.gads) pasākumu izpildi vai pārnešanu, ietveršanu Jaunajā rīcības plānā vides trokšņa samazināšanai Rīgas aglomerācijā laikposmam no 2017. līdz 2022.gadam"* ["Evaluation of Action Plan for Reduction of Environmental Noise in Riga Agglomeration (year 2014) on fulfilling planned tasks and achieving main goals or to carry over the tasks to the new Action Plan for Reduction of Environmental Noise in Riga Agglomeration on period from 2017 to 2022"]. http://mvd.riga.lv/nozares/vides-parvalde/vides-troksnis..

[3] Lācis, M. *"Gains over 60dB."* Author working document.

INDEX

A

B

C

D

E

F

G

H

I

K

L

M

N

O

P

Q

R

S

T

U

V

W